THE CITY

THE NEW URBAN ECONOMICS

URBAN ECONOMICS

THE NEW URBAN ECONOMICS

And Alternatives

H.W. RICHARDSON

LONDON AND NEW YORK

First published in 1977

This edition published in 2007
Routledge
2 Park Square, Milton Park, Abingdon, Oxon, OX14 4RN
Simultaneously published in the USA and Canada by Routledge
711 Third Avenue, New York, NY 10017
Routledge is an imprint of Taylor & Francis Group, an informa business

Transferred to Digital Printing 2007

First issued in paperback 2013

© 1977 Pion Limited

All rights reserved. No part of this book may be reprinted or reproduced or utilized in any form or by any electronic, mechanical, or other means, now known or hereafter invented, including photocopying and recording, or in any information storage or retrieval system, without permission in writing from the publishers.

The publishers have made every effort to contact authors and copyright holders of the works reprinted in the *The City* series. This has not been possible in every case, however, and we would welcome correspondence from those individuals or organisations we have been unable to trace.

These reprints are taken from original copies of each book. In many cases the condition of these originals is not perfect. The publisher has gone to great lengths to ensure the quality of these reprints, but wishes to point out that certain characteristics of the original copies will, of necessity, be apparent in reprints thereof.

British Library Cataloguing in Publication Data
A CIP catalogue record for this book
is available from the British Library

The New Urban Economics

ISBN13: 978-0-415-41825-6 (volume hbk)
ISBN13: 978-0-415-41932-1 (subset)
ISBN13: 978-0-415-41318-3 (set)
ISBN13: 978-0-415-86047-5 (volume pbk)

Routledge Library Editions: The City

the new urban economics: and alternatives

H W Richardson

Pion Limited, 207 Brondesbury Park, London NW2 5JN

© 1977 Pion Limited

All rights reserved. No part of this book may be reproduced in any form by phostostat microfilm or any other means without written permission from the publishers.

ISBN 0 85086 058 X

Printed in Great Britain

Preface

Urban economics is a relatively young field of economics; hardly existing—except perhaps in real estate and land economics curricula—before the 1960s. That decade witnessed a surge of activity in several directions, but especially in relation to urban policies and planning. The theoretical underpinnings remained relatively underdeveloped. Within the last few years, especially after 1971, there has been a growth of interest in urban economic theory, strong enough even to attract the attention of general economic theorists. The models developed were formal and abstract, needing to rely upon very simple assumptions about urban spatial structure in order to derive analytical results. Since the policy inferences of these models were indirect compared with those of the earlier, different type of work, this focus on theory was inevitably associated with a diversion of resources from policy questions, or at least with a dichotomy in the development streams of urban economics.

These new theoretical writings have been named the 'New Urban Economics'—NUE for short. The aim of this monograph is to survey and assess NUE, to evaluate its contribution to urban economics, to offer a few extensions of my own, and to say something about the future direction of the subfield. Also, some alternative approaches to NUE are analyzed. I hope that the book will be of value not only to economists in general and urban economists in particular, but also to other urbanists—planners, urban geographers, regional scientists and others. As this book goes to press, there are probably new papers appearing in this exciting and lively field. A shortcut way of keeping up-to-date is to look at new issues of the *Journal of Urban Economics*, which regularly contains two or three relevant papers in each issue.

I wish to thank: the College of Letters, Arts and Sciences (Division of Social Sciences and Communications) at the University of Southern California, and Dean Donald J Lewis in particular, for providing me with a research grant in the summer of 1975 that enabled me to give time to the writing of the book; Professor Allen Scott, the Editor of this series, for his encouragement and helpful comments; and Mrs Ida Abe for her efficient typing services.

I am also grateful to the Editors of the *Journal of Urban Economics* and *Land Economics* for permission to make use of material published in their journals, and to Lexington Books for permission to draw from a paper published in G J Papageorgiou (Ed.), *Mathematical Land Use Theory* (1976).

Harry W Richardson
School of Urban and Regional Planning, University of Southern California, Los Angeles
January 1976

Contents

1	**What is 'New Urban Economics'?**	1
2	**Antecedents**	
	von Thünen	6
	Hurd	10
	Hoyt's sector theory	11
	The bid-rent function (Alonso)	14
	Muth on urban spatial structure	18
	Wingo on residential land use	22
	A note on rent	25
	A note on utility theory	26
3	**The standard NUE model**	
	The assumptions	31
	The one-dimensional city	31
	Monocentricity	31
	Exclusive zoning	32
	The neglect of production	33
	Housing	34
	Transportation	35
	Homogeneity	36
	The public sector and externalities	37
	Competition in the land market	38
	Long-run equilibrium	40
	Mathematical tools	41
	An example of a simple NUE model	42
	Numerical solutions: the contribution of Mills	45
4	**Implications and extensions of the standard model**	
	The treatment of utility	51
	Transportation land use and congestion	54
	Transportation alternatives—mass transit	58
	Externalities	60
	(a) Air pollution	60
	(b) Public goods	64
	Leisure, family size, and other modifications	66
5	**The monocentric city**	
	Agglomeration economies in the CBD	68
	Empirical evidence of increasing returns	72
	Cities as public goods: CBD substitutes	75
	Exclusive zoning	80
	Employment density gradients	82
	The shape of the transport expenditure function	85

6	**The multicentric city**	
	Introduction	89
	Intraurban hierarchies	93
	Multicentric models	95
	Economics of decentralization	98
7	**More complex residential location patterns**	
	Atypical bid-rent functions	102
	Income and distance	105
	Time constraints and environmental quality	106
	Towards greater realism	109
8	**Locational interdependence**	
	Multiple density gradients	113
	Racial discrimination and rent gradients	117
	Group-preference models	119
	Zoning and land values	122
9	**Towards dynamics**	
	Compromises	124
	Nonspatial models	127
	The dynamics of density gradients	129
10	**An optimum geography**	
	Introduction	131
	The Mirrlees model	132
	Variable production functions	136
	Alternative approaches	138
	Hierarchy models	144
	Space and optimum geography	146
	Conclusion	148
11	**Two residential location models**	
	1 *The possibility of positive rent gradients*	150
	Externality rent	150
	Locational equilibrium	151
	Theory	152
	Conclusion	157
	2 *Discontinuous densities, urban spatial structure, and growth*	157
	Introduction	157
	Density, income, and rents	158
	The basic model	159
	The discontinuous density model	161
	Comparison with the standard neoclassical model	164
	Policy implications	165

12	**Alternatives to NUE**	
	Linear-programming models	168
	Simplicial search algorithms	175
	The possibility of cumulative disequilibrium	177
	The Lowry model	181
	The NBER Urban Simulation Model	185
	The MIT econometric simulation model	194
	The macroeconomic model	194
	Long-run adjustment model	196
	The spatial-allocation model	196
	Evaluation	197
	Spatial interaction in urban models	198
	A spatial-interaction model of residential location	199
	A dynamic urban model	201
	Forrester's *Urban Dynamics*	203
13	**Political economy**	
	1 *Marxism and the city*	213
	Cities and capitalism	214
	Rent	215
	Conflict theories of land-use competition	217
	The role of finance capital	220
	Suburbanization	221
	Engels's views on urban problems	225
	Town versus country	227
	Concluding comments	229
	2 *The unheavenly city*	230
14	**Conclusion: are NUE models operational?**	234
	References	245
	Name index	261
	Subject index	263

1

What is 'New Urban Economics'?

This book focusses on a recent theoretical branch of urban economics, both directly and indirectly (via comparative evaluation of the major alternative approaches to this theory). Mills and MacKinnon (1973) have dubbed this subfield "the new urban economics". As a space saver, this book will use the acronym NUE. This introduction will attempt to distinguish NUE from the rest of urban economics[1].

The boundaries of NUE are very blurred. Some define it very broadly as referring to almost any systematic theorizing about urban spatial structures. Here it is defined rather narrowly as urban economic theories based upon deriving general equilibrium from the principle of utility maximization in a one-dimensional city. According to this view, the partial-equilibrium, utility-maximization models of the early 1960s (Alonso, 1964; Muth, 1961a; 1969) are antecedents of NUE rather than components. Similarly, the more recent work on discrete models (that is, disaggregated two- or three-dimensional cities) by Mills (1972b), Hartwick and Hartwick (1974), MacKinnon (1974), and others falls outside the defined scope[2]. Since the authors of these papers have also written within the narrower field, the delimitation may be a little confusing. However, the boundary is not arbitrary. The linear-programming models characteristic of the discrete-approach school imply a very different philosophical outlook from the utility-maximization models, with less reliance on market processes and much closer links with the urban planning models of the 1960s—links that have implications for the probability and form of empirical testing.

The narrow definition of NUE is not prompted by the desire to set up an internecine fight between NUE and non-NUE models, but is intended as a label for the distinctiveness of its approach. In effect, NUE represents an attempt by economic theorists to explore the usefulness of the methodology and concepts of mainstream economic theory in the analysis of urban problems. In a sense, as several observers and critics have noticed (Anas and Dendronis, Mills and MacKinnon, and Richardson, 1973a), it is closely analogous to aggregate growth theory, the main difference being that its focus is economic behaviour over space rather than through time.

---

[1] There are now two review papers in existence, one by Mills and MacKinnon (1973), the other by Anas and Dendrinos (1976).

[2] Other observers cut the urban economics cake differently. "The hallmark of the new urban economics is the use of fairly sophisticated mathematics—calculus of variations, programming and control theory—to characterize some fundamental aspects of urban structure ... played by almost entirely new players ... general economic theorists who have recently turned their attention to the urban economy" (Mills and MacKinnon, 1973, page 594).

However, in order to contain this analysis within manageable proportions, NUE theorists, like the growth theorists before them, use an artificial vehicle. The 'city' of NUE, like the 'aggregate economy' of growth theory, is much closer to constructs of the imagination than to simplified representations of reality.

To place NUE in historical perspective, a thumbnail sketch of urban economics may be helpful. The intellectual origins of modern urban economics can be traced to a model used for an entirely different purpose—the agricultural location theory of von Thünen (see Hall, 1966). von Thünen's rings for different agricultural crops, where the highest-value crops (in terms of land intensity) are produced closer to the market and there is an inverse relationship between land rent and transport costs, were easily translatable into the concentric zones of urban-rent theory. The description 'von Thünen model' is still used as an approximate term for the standard urban-rent model.

For the first sixty years of this century, most of the writing about urban problems was by either land and real-estate economists (Hurd, Haig, Hoyt, Ratcliff, and others) or by human ecologists and sociologists (Park, Burgess, McKenzie). By inductive reasoning Hurd (1903) stressed the crucial importance of accessibility as a determinant of land values, whereas Haig (1926) highlighted the substitutability between rent and transport costs. Hoyt (1939) provided a link between the land economists and the sociologists by showing the importance of socioeconomic status groups as an influence on urban residential patterns, whereas the Chicago human ecologists demonstrated that social stratification provided an alternative (or, perhaps more accurately, complementary) explanation of residential segregation to income differentials. Although the urban land economists such as Ratcliff (1949) dealt with a whole range of real-estate valuation problems, their work stressed the critical significance of land rent as the instrument for allocating resources within the urban economy. On a different theme, Clark's work (1951) on urban population densities rediscovered the concept of the 'density gradient' that has become so important in urban economic analysis.

At the risk of neglecting other landmarks, it is a fair generalization to say that urban economics 'took off' in the 1960s. One paper prior to 1960 (Beckmann, 1957) pointed the way to the much later NUE models by obtaining a market equilibrium for rents and densities in which rich households live on the urban periphery, although the assumptions used were much more restrictive than are found in more recent models (rent plus commuting costs are a specific function of income, households maximize site size for a given rent expenditure, etc). In a monograph devoted to the measurement of land-value benefits due to transportation improvements, Mohring and Harwitz (1962) employed the simplest form of urban model by assuming Central Business District (CBD) employment and a residential ring in which identical households live on the same lot size.

This implies a horizontal density function, a linearly declining rent function, and (as an equilibrium condition) the complementarity of rent and transport costs.

However, the real origins of modern urban economics were the more or less simultaneous appearance of the first versions of studies by Wingo (1961a), Alonso (1960; 1964) and Muth (1961a; 1969). Since their work is reviewed in chapter 2, only the briefest comments are made here. Wingo developed a market-clearing model of the land market based on simple assumptions about household expenditure, namely that rent plus commuting costs equal a constant sum. This reliance on the older Haig assumption makes Wingo's analysis much less a direct antecedent of NUE than either Alonso's or Muth's. Nevertheless, Wingo made important contributions to the analysis of the influence of transportation technology on urban land use, the significance of which has tended to be neglected. Alonso's work was seminal because it extended urban rent theory by developing a new concept—the bid-rent function—and, perhaps even more important in the context of being a forerunner of NUE, by basing the analysis on the utility function of the individual household (and, since his model also applied to nonresidential land use, on the profit function of the individual firm). Alonso's theory reconciled the concentric-zone model of urban land rent with the utility-maximization behaviour of households (and profit maximization by firms). It only failed to go the whole way because it used a partial rather than a general equilibrium approach.

From the same stable as Alonso (the University of Pennsylvania), Herbert and Stevens (1960) developed a linear-programming model that simulated the residential land market by using an optimizing equivalent of the bid-rent function (that is, maximizing rent-paying ability). This was an important step in that it has been taken up in more recent models as providing a viable, and despite difficulties a more operational, approach to the utility-maximizing model. Muth also used a partial-equilibrium model. but the scope of his analysis was very wide, embracing analysis of population densities and the since neglected question of housing supply. He also discussed many of the complications that follow once the simple assumptions of the standard model are dropped. These include heterogeneous preferences, locally employed workers, and multicentric cities, though his analysis was a passing commentary rather than a development of comprehensive solutions. Muth also subjected his propositions to empirical testing (by using a Chicago data base), and attempted to explore their policy implications.

There was then a hiatus of several years in the development of the theory. Urban economists became interested in other themes, especially policy questions. Whether this was a return to reality or a sidetrack is a question of judgment. In any event, the reason for the diversion was quite clear. In the 1960s governments throughout the world, but especially in the United States, became preoccupied with a wide range of urban problems that were summed up in the phrase 'urban crisis'. These preoccupations resulted in a spate of public policy programmes as an attempt to solve some of these problems. Urban economists responded to public interest in these issues, partly for altruistic reasons—to try to help, partly because topicality provides a degree of self-justification, and partly no doubt to take advantage of the boom in urban research contracts. Although the problems analyzed by the policy-oriented urban economists— such as ghetto strategies, the flight from the central city, metropolitan fiscal problems, urban transportation, poverty and housing, crime, metropolitan consolidation, city size control, and national urbanization policy—have theoretical implications, the theoretical aspects were not seriously discussed. However, the concern with policy issues dissipated towards the end of the decade as the problems were discovered to be more intractable than had been expected, as the research and programme funds dried up, and as governments moved on to struggle with 'crises' of a different kind. Many urban economists still retain a strong interest in policy problems, but most would admit that the steam has gone out of these issues.

Returning to the evolution of NUE, a paper by Strotz (1965) was the first to examine how congestion taxes might be used to convert a competitive spatial equilibrium into an optimum within the framework of a utility-maximization model. But the analysis had little impact at the time and languished in obscurity until the early 1970s. The two papers that began NUE in earnest were by Mills (1967) and Beckmann (1969). Mills examined the role of the production, housing, and transportation sectors within the framework of a general equilibrium model for a city. Although the model was too ambitious for the time and could not be solved properly, it raised many important issues that have been recurrent topics of debate in NUE, such as the role of increasing returns, the conditions of locational equilibrium, the allocation of land to transportation, the nature of traffic congestion, the determinants of the urban boundary, the shape of the rent and density gradients, as well as neglected questions such as the importance of capital-land substitution in urban development. Nevertheless Beckmann's model was the trigger that really started NUE. One factor was that the paper appeared in the first volume of a new journal, the *Journal of Economic Theory*, which has since become one of the most influential vehicles for research in economic theory. Several important NUE papers have appeared in this journal, and the publication of Beckmann's paper there gave the field a respectability among economic theorists that it might otherwise lack. Another reason for the impact of the Beckmann paper was its elegance and simplicity. Although Beckmann was incorrect in certain details, for example, incorrect specification of the boundary conditions (Delson, 1970) and that it is only a special case rather than a general model (Montesano, 1972)[3], his model derived rent and density functions for a log-linear utility function, a linear transportation cost-distance function, and a population described by a Pareto income distribution (but having identical utility functions that included distance as an argument).

By 1970 NUE was on the road. There were several important papers in 1970 and 1971, and the field had developed to the extent that symposia were published in 1972 (in the *Swedish Journal of Economics*) and in 1973 (in the *Bell Journal of Economics and Management Science*). NUE has attracted the attention not only of distinguished spatial economists, such as Beckmann and Mills, but of leading general economic theorists such as Solow and Mirrlees. This book will attempt to evaluate these and other contributions to this fascinating branch of urban economics.

---

[3] Montesano showed that with a zero transport rate the Beckmann model yields the absurd result that residential density and land rent are increasing functions of distance tending to infinity. Also, income decreases with distance only when income is infinitely large at the edge of the CBD. However, Montesano's restatement salvages the model.

2

Antecedents

NUE is not only a very recent branch of urban economics, but it burst on the scene very suddenly in the late 1960s. Nevertheless its origins can be traced back in time to a much earlier date. Indeed, the closest historical antecedent of NUE is the urban version of von Thünen's agricultural land-use model developed in the 1820s. Although some early twentieth century urban-rent analysts (for example Hurd, Hoyt) have had some impact on urban economic theory in general and NUE in particular, the dominant influences have been the models of Alonso, Wingo, and Muth developed in the early 1960s. The contributions of these predecessors to NUE are analyzed in this chapter.

von Thünen

The standard model of urban land use that survives, in modified form, in NUE models derives from the early nineteenth century writings of J H von Thünen (1966). In fact, von Thünen's model refers to the spatial distribution of crops according to yield per unit area around a central town, and it remained for Isard (1956, page 200) to recognize that the model could be applied in an urban context. The CBD substitutes for the central town (or fixed market place) of von Thünen's model, and land uses are arranged concentrically around it with rents and land-use intensity declining with distance. von Thünen's agricultural model has been discussed by Hall (1966), Beckmann (1968), Artle and Varaiya (1974; 1975), Scott (1975c), and Guigou (1972), and Beckmann (1972) has developed an interesting urban version.

von Thünen's own model was abstract, though it was illustrated with numerical examples using data from his own estate (Tellow). His assumptions included: the town is located on a homogeneous plain with land of equal quality; there is no spatial differentiation within the town, hence it can be treated as a single point; landlords maximize profits; they have complete information about production methods, prices, and transport costs; all prices and transport costs are fixed; there are no intermediate goods, and hence no production linkages; and production costs are constant over space. In fact, the model was a little more complicated, in at least two respects. First, his formula for estimating transport costs was based on the assumption that the feed of horses and food for drivers had to be carried along implying a nonlinear transportation function. Second, he assumed that real wages were constant over space, and divided wages up into a 'real'-wage component paid in kind (grain) and a money-wage component. These complications are ignored here.

Considering the most simple example of a von Thünen type model, land rent may be defined as revenue minus costs, and costs can be divided between production costs and transportation costs. Thus,

rent = revenue − production costs − transport costs .

If we start with one sector (crop), and assume that the price is given, the physical output (yield) per unit area is fixed and constant over space, costs are also consistent over space, and the transport function is linear, then

$$p_r = vQ - zQ - Qtr ,\qquad(2.1)$$

where
p_r is the rent at distance r,
Q is the yield per acre,
v is the price per unit of output,
z is the cost per unit of output,
t is the transport cost per unit of distance, and
r is the distance from the market centre (CBD).

Sometimes it is useful to have a price function, and this is easily obtained by rearranging equation (2.1), that is,

$$v = \frac{p_r}{Q} + z + tr .\qquad(2.2)$$

In other words, price is equal to the sum of costs of land, production, and transportation. Given the assumptions of profit maximization and perfect competition (fixed prices and perfect knowledge), all excess profits are whittled away by competition so that production costs include a component for normal profit, and any surplus of revenue over costs is appropriated by landlords in the form of rent.

Equation (2.1) can be simplified still further. Since Q, v, z, and t are all constants, let $x = Q(v-z)$ and $y = Qt$. Thus

$$p_r = x - yr .\qquad(2.3)$$

Differentiating with respect to r, we obtain

$$\frac{dp_r}{dr} = -y ,\qquad(2.4)$$

so that the rent–distance function is negative ($y > 0$), and in this case since y is a constant it is also linear. Since x is also a constant, variations in rent are entirely due to transport-cost differentials. Comparing two locations r_1 and r_2, we have

$$p_{r_1} = x - yr_1 ,\qquad(2.5)$$

and

$$p_{r_2} = x - yr_2 ,\qquad(2.6)$$

and subtracting equation (2.6) from equation (2.5), we obtain

$$\Delta p_{r_1 - r_2} = y(r_2 - r_1) .\qquad(2.7)$$

By introducing two sectors (a and b),
$$p_r^a = Q^a v^a - Q^a z^a - Q^a t^a r \,, \tag{2.8a}$$
$$p_r^b = Q^b v^b - Q^b z^b - Q^b t^b r \,, \tag{2.8b}$$
and simplifying by adding subscripts to equation (2.3)
$$p_r^a = x^a - y^a r \,, \tag{2.9a}$$
$$p_r^b = x^b - y^b r \,. \tag{2.9b}$$

The sector with the highest net yield per unit of area [that is, the highest $Q(v-z)$] uses land most intensively, and will be located in the most central zone. Let this be sector a. It follows from equations (2.9) that, by solving for $r = 0$,
$$p_{max}^a = x^a \,, \tag{2.10a}$$
and
$$p_{max}^b = x^b \,. \tag{2.10b}$$

The maximum potential boundary of each sector's use will be where $p = 0$. Solving equation (2.9a) for $p^a = 0$,
$$r_{max}^a = \frac{x^a}{y^a} \,. \tag{2.11a}$$
Similarly, solving equation (2.9b) for $p^b = 0$,
$$r_{max}^b = \frac{x^b}{y^b} \,. \tag{2.11b}$$

The outer boundary of production occurs when all net yield (surplus over production costs) is absorbed in transport costs. If $(x^a/y^a) > (x^b/y^b)$, sector a will use all the land and sector b will not be produced (cultivated). If this condition does not hold, there will be an inner boundary ($\bar{r}$) where production will be shifted from a to b. This will occur where the two rental functions intersect (that is, where $p^a = p^b$). Setting equation (2.9a) equal to equation (2.9b), and solving for $\bar{r}$, we obtain
$$\bar{r} = \frac{x^b - x^a}{y^b - y^a} \,. \tag{2.12}$$

These findings are shown in figure 2.1. Sector a monopolizes land use where $p^a > p^b$; conversely, sector b dominates where $p^b > p^a$. Thus, the model predicts spatial specialization with no mixed land use (exclusive zoning). The model is easily generalized to n sectors with ranking by distance determined by the highest net yield[4]. If the zones of figure 2.1 are rotated around the origin, the familiar concentric rings are obtained.

[4] If there is a large number of sectors, the rent 'envelope' will appear smooth and differentiable, approximating perhaps to the negative exponential rent gradient of more recent standard theory.

Antecedents

To close the system, market equilibrium needs to be determined, and for this demand functions have to be specified for each product. Let these be $\Phi^a(v^a)$ and $\Phi^b(v^b)$. At equilibrium, supply is equal to demand. Thus

$$\Phi^a(v^a) - Q^a \pi \bar{r}^2 = 0 \; , \tag{2.13a}$$

$$\Phi^b(v^b) - Q^b \pi [(r_{max}^b)^2 - \bar{r}^2] = 0 \; . \tag{2.13b}$$

The model is equally applicable both in rural and in urban land-use situations. Some of the restrictive assumptions, such as constant production costs and linear transport costs, may be relaxed without damage to the basic results. In urban analysis the model is perhaps most easily applied to nonresidential land uses. Superficial problems are created in the residential land-use case. This is because the standard model predicts that the rich live further out than the poor. In terms of figure 2.1, this means that the rich are sector b. At first sight, it is surprising that the poor (sector a) can outbid the rich close to the city centre. However, the simple explanation is that rent, p, is measured in terms of net yield per acre not yield per unit (household). At the city centre $t = 0$, and in the residential land-use model $p^a = (s^{-1})^a(y^a - c^a)$ and $p^b = (s^{-1})^b(y^b - c^b)$, where y is the income, c is the composite consumption, and s^{-1} is the density of households. The poor may bid more for centrally located land because of their tolerance of higher densities. Thus $(s^{-1})^a > (s^{-1})^b$ may more than offset $(y^b - c^b) > (y^a - c^a)$. Since rent is due to savings in transport cost, the rate of decline in the rent gradient will be determined by the increase in travel costs (perhaps including travel time as well as money costs). Transport cost per acre is approximately proportional to the number of people housed on it. Hence, housing types will be arranged in order of decreasing density per area with outward movement from the CBD.

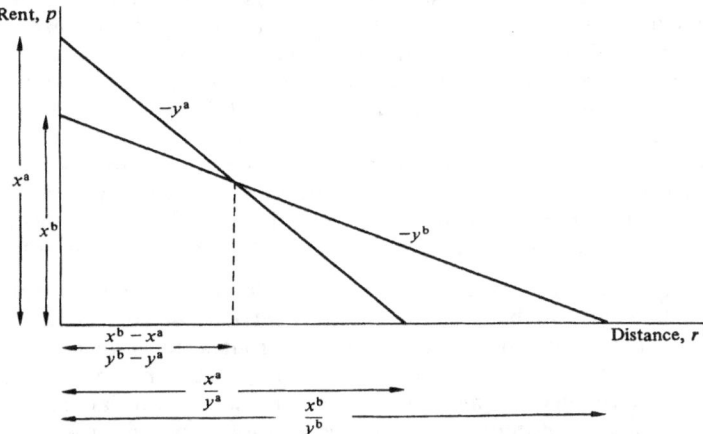

Figure 2.1. Rent function in a two-sector von Thünen model.

Hurd

R M Hurd (1903) is often regarded, with some justification—at least if von Thünen's treatise is considered in its original form as a study of agricultural rent—as being the father of modern, urban land economics. Certainly he emphasized the role of competitive bidding for land in determining urban land uses, and he demonstrated the influence of accessibility on land values—summed up in the frequently quoted sentence: "Since value depends on economic rent, and rent on location, and location on convenience, and convenience on nearness, we may eliminate the intermediate steps and say that value depends on nearness" (Hurd, 1903, page 13). This point has had a notable impact on the form and structure of NUE models.

However, Hurd's analysis went much beyond this simple proposition, and took account of factors still barely touched upon in NUE analyses. For example, he recognized that structures made a huge difference to urban rents. The height of buildings, their layout, and their use are all important. This emphasizes the problems of capital–land substitution and of durability, which are only now beginning to receive attention (see pages 124–129).

His analysis, though verbal, also went beyond the statics that have dominated more recent work. The following quotation neatly summarizes his argument:

"Value by proximity responds to central growth, diminishing in proportion to distance from various centres, while values from accessibility responds to axial growth, diminishing in proportion to absence of transportation facilities. Change occurs not only at the circumference but throughout the whole area of the city, outward growth being due both to pressure from the centre and to aggregation at the edges. All buildings within a city react upon each other, superior and inferior utilities displacing each other in turn. Whatever the size or shape of a city and however great the complexity of its utilities, the order of dependence of one upon another is based on simple principles, all residences seeking attractive surroundings and all businesses seeking its customers.

While the outward glacial movement of a city continues, the daily currents of travel within alter its internal structure. The fluidity of daily traffic shifts utilities, creates plastic conditions in cities and keeps values in a state of unstable equilibrium" (Hurd, 1903, pages 146–148).

Although many of these points are not analyzed in detail in the monograph, their importance is undeniable. The distinction between linear distance and travel time, the importance of transportation routes, the internal changes in spatial structure that accompany urban growth, the conversion of buildings, the influence of environmental quality on residential-location decisions and of market potential on business-location decisions, the interdependence between transportation and land use, and doubts about the stability of urban equilibrium—these are among the very modern ideas implied in this brief quotation. Hurd's insights are truly remarkable for someone writing even before the automobile age.

Hoyt's sector theory

One of the most influential analyses of residential spatial structure has been the radial-sector theory of Homer Hoyt (1939), in effect, a set of inductive generalizations based on empirical study of one hundred and forty-two American cities. Although Hoyt's study is a generally convincing critique of the concentric-zone model as developed by Burgess and Park in the 1920s, it is the latter which forms part of the family tree of NUE models. Hoyt's work has been neglected by modern urban economists for several reasons. His inductive approach led him to develop empirically derived hypotheses without a theoretical framework; or rather, the theory was implicit. Hence, his methodology is not appealing to economists with their love of deductive theorizing. Moreover, to the extent that the sector theory has a base in behavioural theory, this can be developed much more easily in sociological than in economic terms. In addition, Hoyt's explanation of how residential neighbourhoods change takes into account the constraints imposed by existing physical structures; the durability of urban capital is usually ignored by NUE theorists. Finally, an analysis that stresses the heterogeneity of neighbourhoods in different *sectors* of the city has no attraction for theorists who, primarily to simplify the mathematics and to allow them to use their favourite tools of analysis, assume a one-dimensional city.

The residential spatial structure of American cities tends "to conform to a pattern of sectors rather than of concentric circles. The highest rent areas of a city tend to be located in one or more sectors of the city. There is a gradation of rentals downward from these high rental areas in all directions. Intermediate rental areas, or those ranking next to the highest rental areas, adjoin the high rent area on one or more sides, and tend to be located in the same sectors as the high rental areas. Low rent areas occupy other entire sectors of the city from the centre to the periphery. On the outer edge of some of the high rent areas are intermediate rental areas" (Hoyt, 1939, page 76).

Hoyt accepts that these are generalizations subject to qualification. For instance in smaller cities the high-rent area may be adjacent to the business centre. Also, in later papers he argued that there were major differences between US and foreign cities, which called for caution in translating the theory from one institutional context to another. Moreover at any point in time there will be big differences in residential patterns among cities because cities are of different ages and grow at different rates.

The sector theory has a dynamic element connected with the fact that the "high rent neighbourhoods of a city do not skip about at random in the process of movement—they follow a definite path in one or more sectors of the city" (Hoyt, 1939, page 114). The movement of the high-rent area is considered critical to an understanding of the dynamics of urban spatial structure because it tends to pull the growth of the whole city in the same direction. The centre of the high-rent area is the high-rent pole (where the homes of 'the leaders of society' are located). There is a rent gradient around this pole within (and spreading outside) the high-rent area.

Historically the high-grade residential area had its point of origin near the retail and office centre (that is, where the highest income groups worked), which is usually distant from the side of the city where industries and warehouses are located. The critical question was how this high-rent zone moved over time as the city grew. The direction and pattern of its growth is influenced by some of the following factors acting in combination: movement along established lines of communication or towards other existing nuclei; gravitation towards high ground or spread along lake, bay, river, and ocean fronts, provided these are industry-free; expansion towards the open countryside of the city rather than the 'dead end' sections (the lure of golf courses, country clubs, and open fields); gravitation towards the homes of community leaders; pull by the relocation of banks, office buildings, and prestige stores; development along the fastest existing transportation lines; and growth in the same direction over a long period of time. Two qualifications to these influences are the growth of exclusive downtown apartment buildings and the fact that real-estate developers may bend, though not reverse, the direction of high-grade residential growth.

The basic feature of residential spatial structure according to Hoyt is one or two high-rent areas shaped like pie-slice wedges of varying thickness located close to radial lines from the centre to the periphery. The typical high-rent area shifts outwards over time, though at any moment of time it tends to be the most peripheral (that is, the most recently built) area. Low-rent areas tend to be found on the other side of the city (there is mutual repulsion between high- and low-rent areas), and may extend from the centre to the periphery. Intermediate areas may be either next to the high-grade areas or in other parts of the city. Although the gradient in the low-rent area assumes the familiar negative slope, it is much less clear in the case of the high-rent area. There is a negative gradient around the high-rent pole, but the pole shifts outwards over time. This suggests a positive rent–distance function, but with some irregularities (there is a degree of compatibility between this explanation and the externality rent theory discussed below; see pages 150–157). Despite its key role, the high-rent area usually involves less than one-quarter of the peripheral circumference of the city. As the rent peak shifts outwards, the previous peak neighbourhood declines in status. Except in the rare cases of centrally located deluxe apartment areas or 'gentrification' (usually by the middle professional and artistic classes rather than the upper class) of older areas, declining neighbourhoods never recover their former status, though there may be some minor upgrading. If this is so, the growth of the high-rent area must be outwards. Consequently, there is a relationship between neighbourhood quality and age of settlement. There is also a filtering process in which, as the high-rent area shifts over time, middle-class households take over the former high-class areas and these relocated middle class are replaced by lower-class households.

Apart from the mention of community leaders, Hoyt's analysis relies on spatial maps and spatial terms, while areas are classified according to level of rent. Yet is is clear from the above summary that behind the spatial shifts there lies a process of change of social status. The most important aspects of this process are: the social status of neighbourhoods increases with distance from the city centre (the same hypothesis can be derived from the dynamic version of the concentric-zone model associated with Burgess); more precisely, social-status groups cluster within certain sectors, though within each sector there is a zonal pattern of increasing status with outward movement; finally, all areas decline—though often very slowly—in social status over time, owing to the residential mobility of high-status groups into better and newer housing further away from the city centre and their replacement by lower-status households.

The residential location behaviour which leads to these neighbourhood changes is rather complex to examine in the absence of detailed survey data. A simpler test is to investigate the purely spatial hypotheses of the original Hoyt model. Apart from some impressionistic observations, Hoyt's own test of the dynamic elements of his analysis was to look at six cities in three separate years (1900, 1915, and 1936). A recent more satisfactory test is by Richardson *et al.* (1974b). Although this was carried out on a British city (Edinburgh, Scotland), with a housing market very different from the typical American city, the results were encouraging as a support of Hoyt's model. The findings included: the clustering of expensive houses in one or two pie slices; the frequent occurrence of a positive house-price gradient, and where negative the gradient was shallow and became even less steep over time; housing on high ground was more expensive; radial-sector price differentials were large, and the high- and low-price sectors repelled each other; and stability in the high-price sector over long periods of time, though there were occasional jumps from one part of the city to another.

These reasonably consistent results were obtained from a study over sixty-six years (1905–1971) of a city which had certain peculiar characteristics that made it a less than ideal candidate for testing Hoyt. It is topographically eccentric. One-quarter of the housing stock is public, and its development has radically transformed high-status areas. Central-city housing includes some high quality Georgian period housing (part of the New Town) which has retained its status even into the 1960s (Gordon, 1971).

The Scottish land tenure system (feuing) has allowed landowners to place
restrictions on the use of land, including residential density restrictions.
Also, a feature of British housing markets (with the partial exception of
London) is the negligible size of the private rented sector so that the tests
refer to house prices not rental values. However, this should reinforce
social-status effects since owner occupiers are concerned about long-term
capital values which are positively related to neighbourhood status.

Over the years Hoyt's analysis has been subject to a barrage of criticisms
for the following reasons: its spatial determinism (Firey, 1947); the lack
of precision in sector definition; almost total concentration on high-
income groups; neglect of the general structure of the housing stock; a
simplistic view of the social structure, with exaggerated emphasis on
community leaders; abstracting from the effects of planning and
government controls (Rodwin, 1961); and its assumption of a monocentric
city (though Hoyt explicitly discusses the emergence of subcentres).

Although there is a degree of validity in each of these criticisms, the
virtues of Hoyt's approach should not be forgotten, especially in comparison
with the arid elegance of NUE models. First, it breaks away from the
restrictive emphasis on accessibility as the main determinant of residential
location. Second, the emphasis on social status of neighbourhoods as a
key element in house prices and rentals offers insights for understanding
the residential spatial structure of modern cities. Third, the mention of
repulsion both between high- and low-status residential neighbourhoods
and between high-rental housing areas and industry, and the attraction of
prestige business locations for good housing, supports the view that
locational interdependence is important, though this factor is assumed
away in NUE and in most other urban models. Fourth, the complex and
different sectoral patterns revealed in Hoyt's detailed analysis of many
cities highlights the risks involved in reducing the city to a representative
linear ray and in the companion assumption of exclusive zones. Finally,
despite the weaknesses of inductive analysis, Hoyt's careful empirical
studies provide a striking contrast to NUE models which, by making
unrealistic assumptions for the sake of mathematical convenience, may
"conceal the real problems under the masquerade of mathematical symbols"
(Hoyt, 1951, page 262).

The bid-rent function (Alonso)

An important step in the prehistory of NUE models was the concept of a
bid-rent function (originally named the bid-price curve) developed by
Alonso (1964). The most notable feature of Alonso's framework is that it
presents a very partial approach. The problem is conceived in terms of the
location decision of the individual household or firm facing a given rent-
distance function. The rent gradient is not determined within the model,
though, if the bid-rent functions of all potential locators are known,
market equilibrium can be identified by a procedure under which the

distribution of locators with outward movement from the centre is determined by the steepness of their bid-rent functions. Nevertheless, since each locator has a family of bid-rent functions, some coordinates on the rent gradient have to be given exogenously in order to obtain a determinate solution. A second interesting characteristic of Alonso's analysis is that, with minor modifications, the bid-rent concept is applicable to businesses and farms as well as to residences. This gives the analysis a generality absent from many more recent theories.

The assumptions behind the analysis are familiar to those acquainted with the theory of the monocentric city. The city lies on a featureless plain and transportation is possible in every direction. All employment and all goods and services are available only at the city centre. The land market is competitive, free from institutional constraints and the distortions due to existing structures. These assumptions allow residential location to be analyzed in one-dimensional space, and commuting costs to be measured from the home to the centre of the CBD.

The bid-rent function of a household is the set of land rents the household would be willing to pay at various distances from the CBD in order to maintain a constant level of satisfaction, that is, to maintain the same level of utility everywhere which makes the household indifferent between locations. It is derived by maximizing a utility function of the standard type[5]:

$$\max u = u(c, s, r) , \qquad (2.14)$$

where
c is the composite consumption good,
s is the amount of housing space (size of site in Alonso's model), and
r is distance, subject to the budget constraint for a given income, y, of

$$y - vc - p_r s_r - t_r \geq 0 , \qquad (2.15)$$

where
v is the price of composite good,
p is the bid rent, and
t is the transport cost.
The slope of the bid-rent function is given by

$$\frac{dp}{dr} = \frac{v}{s} \frac{u_r}{u_c} - \frac{1}{s} \frac{dt}{dr} , \qquad (2.16)$$

where u_r and u_c are the marginal utilities of distance and consumption respectively. This shows that the bid-rent function slopes downwards. Since $u_r < 0$ because of the nuisance of commuting, and since all the other variables on the right-hand side of equation (2.16) are positive, then $dp/dr < 0$.

[5] Readers with no knowledge of basic microeconomics should at this point turn to the appendix to this chapter for a brief explanation and critique of utility theory.

Another form of equation (2.16) is

$$\frac{u_r}{u_c} = \frac{1}{v}\left(\frac{s\mathrm{d}p}{\mathrm{d}r} + \frac{\mathrm{d}t}{\mathrm{d}r}\right), \tag{2.17}$$

which is a marginal-rate-of-substitution equation with the ratio of marginal utilities on the left-hand side and the ratio of marginal costs on the right. The bracketed term on the right-hand side is the marginal cost of outward movement. Since outward movement produces disutility, marginal costs must be negative to satisfy the constraint of equal utility everywhere. Since travel costs increase with distance, the saving has to be in land costs. With outward movement, cheaper land has an income effect, more than offsetting higher travel costs, and this allows the household to substitute more land and composite consumption for accessibility. Bid-rent functions slope downwards. Lower bid-rent functions imply more utility (because they mean lower land rents). Also, the bid-rent function is single-valued, that is, at any location there will be only one bid rent that maximizes a particular utility level. Finally, bid-rent functions for the same individual do not cross.

To obtain equilibrium, it must be remembered that the bid-rent function is hypothetical, indicating 'willingness to pay' at different distances to achieve constant utility. This implies a family of bid-rent functions for an individual, each corresponding to a particular utility level[6]. The equilibrium location can be determined if the actual rent structure is superimposed on a set of bid-rent functions (see figure 2.2)[7]. The three bid-rent functions (B_1, B_2, B_3) correspond to three different utility levels ($u_1 > u_2 > u_3$). The actual rent structure is given by the

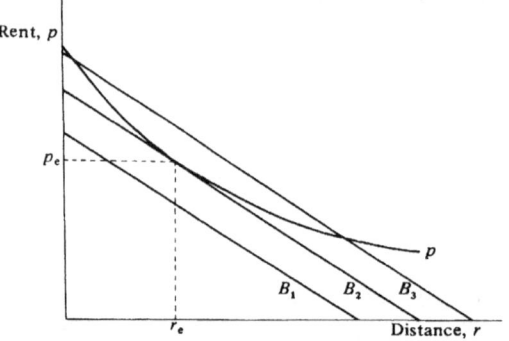

Figure 2.2. Locational equilibrium with bid-rent functions.

[6] The bid-rent functions of different individuals may vary owing to heterogeneous preferences or unequal incomes. See below, pages 20, 36, 37, 67, 102-112.
[7] The bid-rent functions in figure 2.2 are drawn as linear. They could be convex or concave. See Alonso (1964), pages 90-94 and 195-201.

function p. The equilibrium location for the household occurs where the rent function is tangential to the lowest bid-rent function, in this case B_2. The equilibrium location is thus r_e and the equilibrium rent is p_e.

The rent gradient is not obtained directly in the model, though it may be derived in a somewhat clumsy fashion if the bid-rent functions of all residents (or activities) are known, and if one point on the rent function can be fixed *a priori*. This last condition is critical, since otherwise it would not be possible to identify the appropriate bid-rent functions from the sets relating to each locator. The known price reference is frequently the price of marginal land, especially the agricultural land rent at the margin of the city. If this is known, the rent structure can be built up from overlapping bid-rent functions, with the least steep at the margin and the steepest at the city centre. The rent structure thus becomes the envelope of the appropriate bid-rent functions of all users. The envelope is obtained from a chain of pairs of marginal and equilibrium rent-locations. The marginal rent-location of a user of land is the equilibrium rent-location of the user immediately adjacent to him away from the centre. Conversely, the equilibrium rent of a user is equal to the bid rent for that location (marginal rent) of the user next to him but closer to the city centre. However, this procedure may not work if bid-rent functions are not well behaved, in particular, if they do not satisfy the condition of constant ranking by steepness at every point. The solution is much more cumbersome than that of NUE models, though it relies on a similar trick—fixing one point on the rent function in advance, in this case the boundary rent, in some models the rent at the city centre or at the edge of the CBD.

The bid-rent concept is also applicable to nonresidential establishments. In the case of an urban firm the slope of the bid-rent function will be given by

$$\frac{dp}{dr} = \frac{R_r - Z_R R_r - Z_r}{s}, \qquad (2.18)$$

where
s is the site area,
R_r is revenue at distance r,
$Z_R R_r$ is the marginal operating cost arising from change in volume of business R_r, and
Z_r are the marginal increases in operating costs arising directly from outward movement dr.

Thus, the change in bid rent is equal to the change in revenue minus the change in nonland costs, divided by the size of site. Lower bid rents mean higher profits, and equilibrium is obtained by maximizing profits rather than utility. The analysis is based on the assumption that firms benefit from centrality and that revenues fall and costs (apart from land) increase with movement to off-centre locations. This assumption may be relaxed, but the result will be irregular—or even positively sloped—bid-rent functions.

When applied to agriculture the bid-rent function is somewhat different. Because of unlimited entry, farmers are assumed to earn normal profits, regardless of location and the land rent. This means that farmers are indifferent not only to locations but also to rent levels. The relevant bid-rent function is determined, therefore, not by maximization of profits or utility but by the price of the commodity (in the case of the urban firm, this is assumed constant). The rent function for a given market price, on the assumption that produce is shipped to a single central market, is given by

$$p_r^a = Q(v^a - z^a - t_r^a) , \qquad (2.19)$$

where
p_r^a is the agricultural rent at r,
Q is the number of units of crop per unit of land,
v^a is the price at the market,
z^a is the unit production cost, and
t_r^a is the cost of transporting one unit to the market from distance r.
Changes in demand will affect price, and this in turn will influence rent. Competition among farmers ensures that all surplus profits above normal profits accrue to the landlord in the form of rent. The farmer is unaffected provided that he is situated within the rent-yielding hinterland. For a single crop produced at the same production cost everywhere, the actual rent function for a given market price *is* also the bid-rent function. Introducing two or more crops into the analysis leads to an allocation of land use (and in areal terms to concentric rings) among the types of production according to rent-paying ability, with those crops, the output of which yields the larger surpluses of revenues over costs, making the higher bids and hence obtaining the more centrally located sites. This is of course, a simplified version of the von Thünen model, which is also the genesis of modern *urban* land-use theory. This heritage goes far in explaining Alonso's ability to generalize his model.

Muth on urban spatial structure
Arguably still the finest book on urban economics ever written, Muth's monograph on the spatial pattern of urban residential land use (*Cities and Housing*, 1969) is one of the most immediate antecedents to NUE models, though because it was written—and parts were published—between 1959 and 1964 it is of the same vintage as Alonso and Wingo[8]. Although the

[8] In fact, an earlier version of Beckmann's 1969 paper, which started the recent spurt in NUE models, had been circulating in mimeographed form since 1957. One might speculate why, other than imperfect information, the NUE models did not come earlier. Fewer economists (especially those looking around for new fields), less familiarity with the mathematical techniques, and the fact that the love affair between economic theorists and aggregate growth theory was reaching a peak are among the possibilities. The most plausible explanation is that it was not until after 1961 when a number of important publications appeared simultaneously that the field of urban economics (apart from old-style land economics) came into existence.

book is not easy to read because it is so closely argued, it anticipates many of the conclusions of NUE models. Although it does not go as far as some more recent studies in developing an internally consistent general equilibrium model owing to its focus on individual household equilibrium, the housing industry, population distribution patterns and other questions, most of the inferences about the properties of the rent gradient and other findings of NUE models are already there in Muth's work. Moreover he goes beyond recent literature in several important respects. He discusses more exceptions to the standard model (for example, heterogeneous preferences, multicentric cities, and locally employed as well as CBD workers). He is one of the few residential land-use theorists to recognize the importance of examining conditions of housing supply as well as demand. His analysis takes account of greater realism, with explicit consideration given to the influence of the age of buildings, racial segregation, the development of slums, and similar problems. Most important of all, perhaps, is his care to test his qualitative results with empirical evidence, though most of it is obtained from a single city—Chicago.

If there is a limitation to Muth's analysis, it stems from the fact that his Chicago education and experience has rubbed off on him. He is a little too sanguine about the effectiveness of market forces in cities. He downplays the importance of externalities in the urban land market, he argues that the inheritance of physical structures from the past interferes very little with the competitive process, he says very little about the role of planning, and he is not very optimistic about the potential results of public intervention apart from the liberal economist's standard recipe of income subsidies. Although it is arguable that urban economic analysis would benefit from breaking out of its neoclassical shackles, this limitation is not serious from the point of view of studying Muth as a forerunner of NUE since NUE models are all (perhaps with the sole exception of Farhi, 1973) strongly neoclassical.

The Muth study is so rich in its spread and depth of analysis that it is impossible to summarize his arguments satisfactorily in the brief space available here. Apart from mentioning his treatment of household equilibrium—the key element in NUE models and the aspect of urban economic analysis stressed by Muth's contemporaries—attention will be focussed on problems neglected in the more recent work.

The treatment of household equilibrium is familiar to those acquainted with the NUE literature. A household, with one member working in the CBD, maximizes its utility function that contains housing and all other goods and services, subject to a budget constraint that expenditure on composite consumption, housing, and travel should not exceed total income. The equilibrium conditions are: the marginal utilities per unit of expenditure on housing and consumption must be equal; and, for locational equilibrium, the marginal change in housing expenditures with a change in location must be equal to the marginal change in travel costs.

If marginal travel costs are positive, equilibrium requires that house prices decline with distance and that housing consumption increases with distance. If marginal travel costs do not increase with distance and if the price elasticity of demand for housing is not greater than unity, house prices decline with distance at a nonincreasing rate.

The influence of income differences on location cannot be predicted *a priori*. Higher incomes increase housing expenditures favouring a less central location, but, in the case of earned income, longer commuting trips are more costly to the wealthy because the value of travel time is related to income. However, empirical estimates suggest that higher incomes will increase housing expenditures by relatively more than marginal commuting costs, with the result that higher-income worker households tend to live further away from the CBD. In studying the determinants of decentralization, other factors apart from income changes have to be taken into account. Declining transport costs, for instance, increase the equilibrium distance from the CBD. Increases in house prices, on the other hand, tend to reduce the equilibrium distance. Stronger preferences for space rather than structures, the preference for single-family dwellings, and incentives to home ownership also reinforce decentralization. Muth's analysis allows the restriction of all employment in the CBD to be dropped without disturbing the results of his model, provided that there is a negative wage gradient. This is a rare reference to an interesting phenomenon, analyzed by Moses (1962).

Housing supply considerations have also to be studied in attempts to explain decentralization. One of the most innovative aspects of the study is the analysis of equilibrium conditions among housing producers, and their implications for residential land use. This question has been neglected in NUE models. Indeed a feature of these models is the failure to draw a distinction between the demand for housing and the demand for land. Muth shows that the decline of housing prices with distance (a major implication of the household equilibrium conditions) implies that the value of residential land must also decline if housebuilders are to make the same profits everywhere (a condition of locational equilibrium in the housing industry). Moreover, land values decline with distance much faster than house prices, and this induces builders to use more land relative to other factors of production at greater distances from the city centre. This is a major factor in the decline of population densities with distance.

Two important factors relevant to explanations of the decline in the intensity of housing output with distance are the nature of housing production functions, and variations in the fraction of land area used for residential purposes. The empirically observed rise in the share of the value of housing output attributable to land, coinciding with an increase in land values relative to construction costs, is consistent with the assumption that all builders have the same Cobb–Douglas production function.

Solutions are: an increase in the land exponent due to nonneutral
technical change or to the introduction of different house types with more
large-lot dwellings; or the assumption that builders all have the same
production function but that the elasticity of substitution between land
and nonland factors is less than unity. The latter explanation was favoured
by the lack of evidence for a positively-curved, log-linear function of land
values relative to distance and by the tendency for the residential population
to become more decentralized with increasing city size. The fraction of
land devoted to residences may be expected to increase with distance
owing to the price elasticity of housing supply increasing with distance.
Empirically, however, Muth found that in Chicago the residential land
fraction was not very responsive to distance, perhaps because of the
speculative withholding of land from development. He concludes that the
simpler assumption of a constant residential land fraction is perhaps not
too distorting. Of course, if housing supply is more elastic at greater
distances, this is yet another determinant of more rapid population growth
in the suburbs.

Muth also introduces the concept of "complete uniqueness"—the fact
that land parcels and housing services may differ for reasons not connected
with location (in the sense of distance from the CBD). Urban land is
heterogeneous and subject to indivisibilities primarily because of the
existence of fixed capital improvements such as buildings, streets, and
utilities, but also because of diversity of ownership of contiguous parcels.
The relevant factor of production then becomes real estate rather than
land. The result of this bow in the direction of realism is that there is
much greater variety in the price of housing and in land values than is
implied in the simple distance functions. Since equilibrium in the land
market still requires that each parcel be devoted to its most profitable use,
complete uniqueness opens the door to conversions from one land use to
another. Nevertheless, despite the complications, this relaxation does not
alter the basic spatial structure very much. If house prices decline with
distance, then the value of housing output per parcel of real estate in the
complete uniqueness case also declines with distance in a similar way.

The treatment of the distribution of population density is familiar but
interesting. Muth shows how his model generates a negative exponential
density gradient of the kind observed empirically by Clark (1951) some
years before. He also considers the influence of secondary centres of
employment, shopping centres, and rapid transit routes on population
distribution patterns. These modifications, generally ignored in NUE
models, produce easily predictable irregularities in the density distribution,
and generally tend to flatten the density gradient.

The discussion of racial housing segregation in the study is also rarely
mentioned in NUE models, though it has been analyzed quite frequently
by more traditional urban economists. Muth criticizes the view that
segregated housing markets are based upon discrimination (for example,

aversion or conspiracy theories), and argues that if blacks pay more for equivalent housing than whites (the Chicago evidence suggested that the differential was small) this is to be explained by faster growth in black demand for housing. Also, following Bailey (1959), he develops a spatial model of dual housing markets which shows that if house prices in the interior of the markets are the same both in the black and the white areas, and if whites have a greater aversion to living among blacks than *vice versa*, prices on the white side of the boundary will be lower than on the black side. The black area would tend to expand, and—assuming stable demand—house prices in the interior of the black area would drop. If they do not, the explanation must be one of rapidly increasing demand.

Muth attacks most of the contemporary explanations of the growth of slums. These usually stress supply factors in the form of an increase in supply resulting from a decline in demand. This, in turn, is explained by many factors: the growth in automobile use, physical obsolescence, poor planning, increasing incomes, inadequate public services, external economies and other market imperfections, and the effects of property taxation. Muth believes that these explanations fail to account for three empirical facts: slum housing is relatively expensive; urban renewal projects are unprofitable; and the decline in the proportion of substandard housing in the United States cities. A preferable explanation which is consistent with these facts is that the growth in slums is due to an increase in demand for them which reflects an increase in low-income population in the cities. Thus, slums are primarily the result of the poverty of slum inhabitants. Public policies such as urban renewal, rapid-transit subsidies, and suburban development strategies are likely to make matters worse. If the slum problem is to be attacked effectively, the point of attack must be its direct source—the incomes of slum residents.

The observations in the last two paragraphs go beyond Muth's model of residential spatial structure. The theoretical content, though still present, is overshadowed by the empirical analysis and the policy inferences. Its value is primarily to show that Muth, unlike most of the NUE modellers, recognizes that understanding residential land use requires a broader approach than that of abstract models. The blend of theoretical analysis and empirical testing has rarely been achieved in urban economics. Muth's monograph raises issues which have not yet been included within the scope of NUE models.

Wingo on residential land use

Wingo (1961a) is the third of the triumvirate of pre-NUE modellers. His analysis differs from the others in two major respects. First, he gives more attention to the role of transport, and uses a more complex transportation function. Second, his treatment of household equilibrium is simpler, since he does not employ the concept of a utility function with substitution possibilities among different forms of expenditure. Instead he keeps the much older assumption (associated with R M Haig, 1926) of the complementarity between rent and transport costs. In particular, the sum of these is assumed to be a constant, and this is a special case of the more general utility-maximizing model in which income and spending on composite consumption are assumed to remain constant.

The model is based on the standard assumptions of the single-CBD city, and the population has the same income and tastes. Transport technology, the marginal value of leisure, and the marginal value of residential space are given. Prices, apart from the price of urban land, are assumed constant. Transport costs consist of money and time costs; no allowance is made for the disutility and inconvenience of travel. Money costs include distance costs and terminal costs which depend on CBD congestion. The time cost is the commuting time to work, priced according to the marginal value of leisure. The rent for a site is simply the unit price multiplied by the size of site occupied. Rent is assumed to be zero at the urban boundary (that is, it excludes the opportunity costs of urban land). Given the assumption that rent plus transport costs equal a constant, this means that the value of the constant is transport costs between the city centre and the urban boundary. Also, since transport costs increase with distance, rents must decline with distance. Rent is thus a payment for accessibility, hence Wingo's term "position rent" defined as "the annual savings in transportation costs compared to the highest cost location in use".

Transport cost at the urban boundary, located at distance $\bar{r}$, is equal to $t_{\bar{r}}$. Rent at any other location ($P = p_r s_r$) is equal to $t_{\bar{r}}$ *minus* transport costs actually incurred (t_r), thus

$$p_r s_r + t_r = t_{\bar{r}} = \text{a constant} ,\qquad(2.20)$$

or

$$p_r s_r = t_{\bar{r}} - t_r .\qquad(2.21)$$

It is assumed that the space-demand function is constant among households, and has the usual property of being downward sloping from left to right. Households prefer more land to less land, but the higher the price the smaller the space consumed. More specifically, Wingo assumes a log-linear demand function,

$$s_r = \left(\frac{\alpha}{p_r}\right)^{1/\beta} ,\qquad(2.22)$$

where α and β are parameters and $\beta > 1$. These two equations, (2.21) and (2.22), determine household equilibrium when solved simultaneously. The first equation shows the value of position rent, whereas the second determines the site size and the price of land.

In the transition from individual to market equilibrium, the first point to be noticed is that the demand function determines the population density gradient. Since $p = Ps^{-1}$ (where s^{-1} measures population density, assuming one person households), equation (2.22) may be rewritten as

$$s^{-1} = \left(\frac{p}{\alpha}\right)^{1/(\beta-1)}. \tag{2.23}$$

Since the demand curve is negatively sloped, and the position rent gradient is negative, it follows that the density gradient is negative. This finding is virtually unanimous in urban residential land-use models.

The density gradient and the demand function determine the demand for land at each distance from the city centre. Market equilibrium requires equating the demand for and the supply of land. In a circular city in which all the land is available for residences, the supply-of-land function is

$$\sum_{r=1}^{\bar{r}} S_r = \pi \bar{r}^2, \tag{2.24}$$

where S_r is the supply of land at each distance. Total population, N, is obtained by integrating its density, s_r^{-1}, over the area of the city, thus

$$N = 2\pi \int_0^{\bar{r}} r s_r^{-1} dr. \tag{2.25}$$

By solving equations (2.23), (2.24), and (2.25) simultaneously, $\bar{r}$, p_r, and s_r are obtained, and this is the solution to the model.

There is a serious difficulty with this model; it is a consequence of equation (2.20). The constant-expenditure constraint on rent plus transport costs, combined with the assumptions of identical incomes and constant expenditures on composite consumption, means that every household has the same demand curve for land regardless of location. The fixed budget implies that with outward movement, and as transport costs increase, total rent expenditures must fall. The price of land declines with distance, partly because central land is more valuable because of its greater accessibility, partly because the supply of land increases with outward movement in a circular city, whereas the number of demanders tends to fall as density declines. If a fall in the price of land is associated with declining housing expenditures, this means of course that the price elasticity of demand is greater than -1. This is a condition of equilibrium in Wingo's model. If the price elasticity of demand were equal to -1 [a plausible value according to Muth (1969) and others], there would be no solution. If the price elasticity is less than -1, equilibrium is possible only if the rent gradient is positive. This restriction makes the model weaker than those based on utility-maximization criteria.

However, the value of Wingo's monograph does not rest solely on the development of a somewhat flawed model of residential spatial structure. More than one-half of his research deals with the role of transportation in the spatial structure, and is concerned with such issues as transportation technology and the economics of the journey-to-work. In particular, he presents one of the earliest discussions of congestion and of the value of travel time. His comparative analysis also shows how different preferences for space and time (accessibility) lead to a stratified (concentric zone) pattern of residential land use and he explores the impact of the growth of secondary employment centres on the rent and density gradients. Finally he analyzes the influence of changes in public policy, especially with respect to transportation. Although the method of analysis is less rigorous than in the NUE models, the Wingo study is broader in scope and arguably richer in insights.

A note on rent
NUE models do not provide an overall view on the nature of urban land rent. An obvious support of this statement is the failure of NUE to say anything about nonresidential rent (almost always treated as exogenous) or, *a fortiori*, about industrial rent. Less apparent, though equally significant, is that NUE theories, and the Alonso-Wingo-Muth heritage from which they evolved, lie outside the mainstream of land-rent theory. In classical theory land is an input into the production process analogous to labour and capital. Ricardian differential rent is a transfer payment to landowners made possible by the fact that more fertile land allows commodities to be produced at lower costs. In Marxian theory land has no value, but has a price because the institution of private property allows landowners to extract part of the total surplus value produced by labour. In a way this might be reconciled with NUE by arguing that rental of urban space is the price of admission to the urban labour market, and competition for space enables landlords to appropriate some of the benefits of higher urban productivity [this view is shared by radicals such as Edel (1972) and NUE theorists such as Mirrlees (1972) and Barr (1972)].

In fact, the approaches of both Ricardo and Marx were nonspatial, whereas the spatial aspects of rent theory are traceable to von Thünen, and before him to Sir James Steuart (Scott, 1976b). Although von Thünen's model was phrased in terms of producing and distributing commodities, his primary emphasis was on the concept of accessibility as the determinant of rent. Some sites are more valuable than others because they offer greater accessibility to key locations (the market, the workplace, etc). The *minimum* value of this accessibility is the transport cost savings that are gained from a more proximate location. Land is not valued for its own sake (though as incomes rise there is a conspicuous consumption element to be gained from occupying a larger residential site), but because of its location. High rents are a surrogate for the benefits of accessibility. It is for this reason that NUE modellers have focussed on residential space *per se* rather than on housing production, a sector that uses land as an important input (Borukhov, 1973; Mills, 1972a, are rare exceptions to this generalization).

Appendix

A note on utility theory

For urban analysts with a negligible economics background it may be helpful to describe utility theory, since NUE models are based upon the assumption of utility-maximizing behaviour by households. Utility theory is the standard approach used by economists to explain how consumers make rational choice decisions. Consumers are assumed to maximize their utility (or welfare) subject to constraints on their choice, the most obvious of which is income. Their utility is assumed to be a function of the amounts of different goods and services they consume. In the simple two-goods case, this may be represented as

$$\text{maximize } U = U(a, b), \quad (A2.1)$$

$$\text{subject to } y = p_a a + p_b b, \quad (A2.2)$$

where
U is utility,
a is the quantity of good a,
b is the quantity of good b,
p_a is the price of a,
p_b is the price of b, and
y is income.

The simplest way of illustrating how the problem is solved is via indifference curves. It is assumed that the consumer knows his own tastes, that his tastes are stable and consistent, and that commodities are divisible. These assumptions enable us to map the individual's preferences and to represent combinations of 'a' and 'b' that are equivalent to him (that is, combinations between which he is indifferent—in other words, which confer equal utility). These equivalent combinations may be plotted (conceptually at least) on an indifference curve. A preference map for goods 'a' and 'b' may be represented in two-dimensional space. In this space there will be a great many indifference curves, of which three (I_1, I_2, I_3) are shown in figure A2.1. If goods are divisible, the curves will be continuous. If both goods are desirable, the curves have a negative slope because if one combination has more of 'a' than another this means less 'b' for the combinations to be equivalent. Also, it follows that consumers prefer higher to lower indifference curves, since these imply more of both commodities (or at least more of one and the same amount of the other).

It is usually assumed that the indifference curves are convex to the origin. The simplest way of ensuring this is to assume that consumers buy some of each commodity (if only one commodity is bought, this implies linear or concave indifference curves). The more substantial explanation is, however, based upon the principle of diminishing marginal utility. Marginal utility is the increase in total utility divided by the increase in the quantity of the commodity, and the argument is that marginal utility

Antecedents

will decline as the number of units consumed increases. This implies convex indifference curves. If u_a is the marginal utility of 'a', and u_b is the marginal utility of 'b', then for indifference to hold on an indifference curve

$$da u_a + db u_b = 0 . \tag{A2.3}$$

The slope of the indifference curve, called the marginal rate of substitution of 'b' for 'a', is obtained by rearranging equation (A2.3), thus

$$M_{ba} = -\frac{db}{da} = \frac{u_a}{u_b} . \tag{A2.4}$$

where M is the marginal rate of substitution.

Moving to the left on the indifference curve, u_a gets larger and u_b becomes smaller if the principle of diminishing marginal utility holds. Thus the slope $(-db/da)$ of the indifference curve becomes greater, that is, it is convex. One problem (mentioned below) is that diminishing marginal utility guarantees convexity only if another assumption is made—that utilities of different commodities are additive. If utility combines multiplicatively (that is, if the utility of one good depends upon the consumption of others), the indifference curves may not be convex.

Indifference curves reflect subjective preferences. To determine which combinations of goods will be chosen we have to introduce the objective world in the form of relative prices and the income constraint. If we assume constant relative prices, the price line will be linear with the slope $-p_a/p_b$. The relevant price line is the consumer's budget constraint. On the assumption he spends all his income [that is, assuming equation (A2.2) is an equality] then from equation (A2.2)

$$b = \frac{y}{p_b} - \frac{p_a}{p_b} a . \tag{A2.5}$$

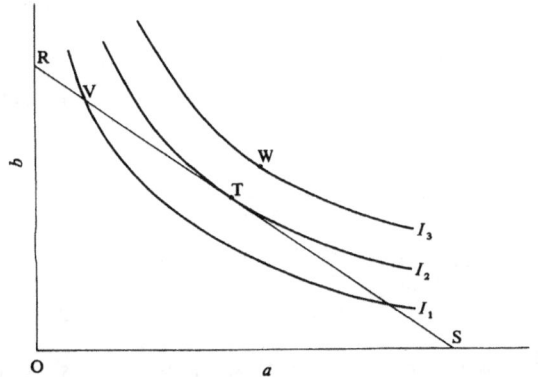

Figure A2.1. Utility maximization with indifference curves.

In figure A2.1, where RS is taken as the budget constraint for a given income y, all the possible combinations of 'a' and 'b' open to the consumer lie in the triangle RSO, and if all income is spent the possible combinations lie on the budget line RS.

It is now not difficult to show what utility maximization means. To maximize utility the consumer will select from the feasible combinations of goods that which is on the highest indifference curve. In figure A2.1, given the budget constraint RS, the individual will select combination T on indifference curve I_2. It is superior to combination V on indifference curve I_1 and to any other combination on indifference curves below I_2. It is inferior to combination W on I_3, but that is infeasible because it is beyond the consumer's income. Thus the utility-maximizing combination is found where the budget line is tangential to the highest indifference curve, that is, at T. At that point the budget (price) line and the indifference curve have the same slope. Thus the maximum utility condition is

$$M_{ba} = \frac{p_a}{p_b}, \tag{A2.6a}$$

or

$$\frac{u_a}{u_b} = \frac{p_a}{p_b}. \tag{A2.6b}$$

Rearrangement yields the standard first-order maximization condition that the marginal utility/price ratios of both goods must be equal. The first-order maximization condition enables an extreme solution (either a maximum or a minimum) to be obtained. To obtain a maximum, a second-order condition is needed, namely to ensure that the indifference curve is convex. The condition of convexity is

$$\frac{d(M_{ba})}{da} < 0, \tag{A2.7a}$$

or

$$\frac{d(u_a/u_b)}{da} = \frac{u_b^2 u_{aa} - 2u_a u_b u_{ab} + u_a^2 u_{bb}}{u_b} < 0, \tag{A2.7b}$$

where

$$u_{aa} = \frac{\partial^2 U}{\partial a^2}; \quad u_{bb} = \frac{\partial^2 U}{\partial b^2}; \quad \text{and} \quad u_{ab} = \frac{\partial^2 U}{\partial a \partial b}.$$

Diminishing marginal utility ($u_{aa} < 0$, $u_{bb} < 0$) is not sufficient to ensure that the left-hand side of equation (A2.7b) is negative. The assumption that u_{ab} is zero, that is, the additive utility assumption, must also be made.

There have been numerous objections to utility theory, and these should be mentioned. One of the most critical is the argument that utility is not directly measurable, except by crude ranking of preferences. This is described by saying that utility is ordinal rather than cardinal. The consequence is that the indifference-curve analysis above can no longer be discussed in terms of diminishing marginal utility, only in terms of the marginal rate of substitution. Utility becomes tautological referring to whatever consumers maximize. Consumers are still rational, in the sense that they try to do their best for themselves with what they have, and this may be described as maximizing utility, but utility is no longer an objectively measurable concept and utility theory cannot serve as the basis for an operational model. The theory becomes trivial and empty.

Another problem is that the world is uncertain and life is risky. Since income can be consumed now or saved, and spending may be deferred, individuals maximize *expected* utilities, even if utility can be measured. This implies that different probabilities must be assigned to alternative outcomes, and hence a serious element of indeterminacy is introduced.

A further related criticism is the hypothesis that consumers do not maximize anything, but 'satisfice', that is, they are content with a satisfactory outcome. This is a common idea in the theory of the firm. Less frequently accounted in the theory of consumer behaviour, a similar idea has been analyzed at length by Devletoglou (1971). He argues that behaviourally individuals may be indifferent between two outcomes, even though one is 'better' than the other, if the difference between expected utilities is small enough. In other words, there is a threshold band of indifference. This hypothesis strikes a blow against utility maximization, particularly because it destroys the key transitivity assumption. For instance let m be the threshold band of indifference measured in utility terms, and assume that an individual has to choose from three alternatives with expected utilities U_1, U_2, and U_3, where $U_1 > U_2 > U_3$. Assume that $(U_1 - U_2) < m$, $(U_2 - U_3) < m$, but $(U_1 - U_3) > m$. Then the consumer is indifferent between 1 and 2 and between 2 and 3, but he will prefer 1 to 3. Devletoglou's applications of his theory include spatial contexts, especially locational duopoly (two-seller) models.

There is a case for applying these ideas in the residential housing market. A residential locator is not only subject to income and other resource constraints (for example, the availability of mortgage funds for a housebuyer) but also to time and space constraints. He has to find a home within a given time period, and he will have to restrict the spatial limits of his search field. The costs (especially the time costs) of obtaining market information may be substantial and uncertain. Only a small fraction of the housing stock will be available for rent or purchase in any interval of time, and the ideal location or ideal structure for a particular locator may not be obtainable within the time allotted for the search. All these considerations suggest that the search for the 'best' rather than a 'satisfactory' house and location may be fruitless and, more important, costly. Some houseseekers live in expensive temporary accommodation and house prices and rent may be increasing over time. The pressures to choose sooner rather than later may be very great. Whether one describes this kind of behaviour as 'satisficing' or as maximizing subject to severe additional constraints, especially of time, is a matter of semantics or of taste. Devletoglou's threshold band of indifference is merely another way of saying that the transaction costs of seeking out marginally superior outcomes may be too heavy. The response of 'it's not worth the trouble' is not on this view incompatible with the concept of a utility maximizer.

Finally, adoption of utility theory implies that real consumer choice exists. The old epigram about the rich and the poor having the same freedom to steal bread or to sleep under bridges is very relevant here. The use of the budget equation implies that all transactions are voluntary. But the radicals argue that this assumption may not hold, especially in the housing market. Housing is a necessity, and it is indivisible (an infringement of the assumption needed to derive continuous indifference curves and budget lines). A low-income household may have virtually no choice at all as far as the housing market is concerned. Low incomes and indispensable housing needs on the demand side, a possible scarcity of low-cost housing and strong market power exerted by slum landlords on the supply side—this combination may force the poor household to take any housing it can get at high rents. The compulsion may absorb a very high proportion of the household's low income, leaving an inadequate residual for allocation among alternative priority spending claims. In effect, the budget line between housing and 'all other goods and services', that is, composite consumption, reduces to a point. Thus the radicals would argue that utility theory with its pretensions of relatively free consumer choice is an ideological sham, and could only operate—if at all—under conditions where the distribution of income was relatively equal (see chapter 13 for an elaboration of the radical perspective).

3

The standard NUE model

The assumptions
Much insight can be gleaned about the nature and characteristics of NUE models by examining their basic assumptions. However, an examination of NUE assumptions is fraught with difficulties because the assumptions vary somewhat from model to model. In particular, there has been evolution in NUE models with the assumptions becoming less restrictive over time. The most satisfactory way of proceeding is to discuss the assumptions of what might be called the 'standard' or the 'basic' model, taking care to mention where there have been significant relaxations in the assumptions in later models.

The one-dimensional city
NUE models assume a circular city (with the possibility of pie-slice radians taken out for topographical constraints) in which transportation is possible in all directions. This means that the city can be represented by a linear ray from the CBD to the urban boundary. In other words, the city deals with one-dimensional space rather than discrete areal zones. The main reason for this simplification is mathematical. NUE modellers use calculus as their tool of analysis, and hence they require distance functions to be smooth and differentiable. This requirement, incidentally, also rules out the possibility of discontinuities in rent and density gradients (for arguments why models predicting discontinuities might be helpful see chapter 11, pages 157–167).

Monocentricity
The assumption that the city has a single centre, the CBD, is a corollary of the one-dimensional city assumption. Secondary centres would appear as peaks on the rent functions, a fact that complicates the mathematics a little. More seriously, unless the secondary 'centres' take the shape of annular rings, they cannot be handled via a one-dimensional analysis because the subcentres will be located in some sections but not in others; thus, the city could not be represented by a single linear ray. The advantages of the monocentric assumption are, of course, considerable. For instance, if all jobs are centralized the journey-to-work of each resident may be exactly specified if his residence is known. The one centre is also assumed to be centrally located. In fact shifting the CBD off centre would also, because of the urban spatial structure implied by NUE models, shift the city as a whole in the same direction. The only exception would be the presence of insuperable topographical obstacles such as mountains, rivers, lakes, and the ocean. Interestingly, Beckmann (1976) has shown that if the CBD is assumed away as a workplace and some other reason for agglomeration is substituted (for example, a desire

for social interaction), rent and density gradients still develop around the new centre ('piazza' or 'plaza') though the peak is rounded rather than a sharp point (see pages 78-80).

There have been a few relaxations of the monocentricity assumption. Papageorgiou and Casetti (1971) and Papageorgiou (1974) have explored the implications for rents and densities in a regional urban hierarchy context. They found that local rent maxima do occur, but additional restrictive assumptions were needed to obtain a completely determinate solution (see pages 95-98). Lave (1974) examined the economics of decentralization, but only from the narrow point of view of trading off the benefits of economies of scale in production for lower rents and freight costs (see pages 98-101).

Exclusive zoning
The standard NUE model assumes that the city is a doughnut. The hole in the middle is the CBD in which all production takes place (it really is a 'hole' because the CBD has never been properly analyzed, despite the researches by Livesey, 1973, and others). The doughnut itself is a residential ring in which land is devoted only to housing, apart from the land used for transporting commuters to and from the CBD. There are no jobs outside the CBD (not even in building, construction, and civil engineering) and no homes inside the CBD. This 'exclusive zoning' rules out the possibility of competition between nonresidential land uses (apart from transportation) and residential land use, and it eliminates the complication of nonradial, non-CBD work trips that would not only mess up the measurement of transportation demand but would also disturb the smoothness of the rent function. Exclusive zoning means that NUE provides, in effect, a model of the suburbs not of the city as a whole.

There is clearly an inconsistency between the theoretical desire for simplification and the strong empirical evidence of increasing employment decentralization. "It is somewhat ironic that economists are becoming proficient at building models in which all employment is in the CBD and all housing outside it, when cities look less and less like that paradigm" (Mills and MacKinnon, 1973, page 600). However, it is even doubtful whether the paradigm was ever valid, since even nineteenth century cities exhibited mixed patterns of land use (Fales and Moses, 1972). Of course, it is a question of judgment as to whether the violence to the facts is justified as a theoretical simplification. Much depends on the implications of the theory, such as the lack of competition for land between housing on the one hand and commerce and industry on the other. To evaluate whether or not this is important requires an understanding of how the land market operates. On this point, as on others, the perception of one analyst will differ from that of the next.

The neglect of production
At best, NUE models contain three sectors—housing, transportation, and production. The first two receive most of the attention (though still limited), and are discussed below. In addition agriculture is sometimes brought into the model as an aid to determining the urban-rural boundary (for example, Mirrlees, 1972). Production is treated as a single sector producing a composite consumption good within the CBD. In many early models the CBD was treated as a point so that production used no land. Later the CBD was assumed to have a finite (usually fixed) radius, but CBD employment was determined exogenously. In other models the production assumptions were that one commodity was produced under constant returns to scale. Capital was frequently subsumed under nonland inputs, yet the capital-labour ratio has a vital impact on urban spatial structure (for example, as a determinant of transportation demand). The CBD was treated as an 'empty box' and the production sector as having no intrinsic interest.

The weaknesses of these approaches are self-evident. Possibly the primary reason for the existence of cities (or at least for their growth) is increasing returns to scale from contiguous production (of both like and unlike activities). The extent of these increasing returns is the chief determinant of city size. Later NUE models have attempted to cope with this issue. Dixit (1973) was the first to consider it in detail via analysis of the case of a production function for a homogeneous consumption good by using land and labour inputs (no capital) subject to increasing returns. He admitted two drawbacks: the neglect of multiple products and hence of interurban trade; the assumption that *all* CBD land is used for production. The latter limitation implies that once commuters cross the CBD they are at work (in Dixit's model, in the factory). The introduction of multiple products not only has implications for interurban trade, and hence for freight transport, but also for the location of production. Dixit's factory town is unrealistic because few large cities have factories in their CBDs. If two production sectors were permitted, it might be reasonable to assume that one was a manufactured commodity and the other a service. It would then be sensible to permit the service to be supplied within the CBD, whereas the manufacturing plant could be located at the edge of the city, on the periphery beyond the residential ring. A model of this kind has yet to be developed by NUE. Although it would be more complicated, since there would be two-way commuting flows, the difficulties would not appear insuperable—especially if some correspondence was assumed between occupational and residential stratification.

Several NUE economists have analyzed land use for transport within the CBD (Mills and de Ferranti, 1971; Sheshinski, 1973; Livesey, 1973), but for the exogenous employment case. In later work Livesey (1976) and Alao (1976) have relaxed this latter assumption and have shown that one result is a land-rent gradient *within* the CBD. Despite this progress, neither the production sector nor the CBD has been given the same attention in NUE models as housing and transportation in the residential ring.

Housing
The last statement might suggest that housing has been analyzed to death by NUE theorists. On the contrary, they have treated housing in an exceedingly simple way. The standard assumption is that the demand for housing is a derived demand for land, and that it is the demand for land that counts. If land is homogeneous (and there are no environmental externalities), there are merely two dimensions to housing demand: plot size and location, where, because of the monocentricity and one-dimensional space assumptions, location simply means distance from the CBD. Moreover, since the density gradient is unique for any specific set of parameters, choice of a site size also implies choosing a specific location, or *vice versa*. The two dimensions of housing choice are jointly determined. In many NUE models this approach has yielded the simplifying benefit that the household utility function need not contain distance or transport costs, since these are implicitly taken into account in the amount of housing consumption.

The assumption that housing demand is merely a demand for urban land is as if urban residents have been given tents by a relief organization and they pitch them at the site size (and location) that maximizes their utility. This analogy would also be helpful to cope with the situation where changing circumstances required locational changes to restore equilibrium, since locational inertia is minimized with an analogy of this kind. The neglect of housing is understandable. Since housing cannot be a centralized industry (even factory-made units require on-site assembly), the introduction of housing supply (Muth, 1969; Mills, 1972a; and Oron *et al.*, 1973, have all explicitly included a housing-supply sector, though Muth alone has allowed for intraurban variations in construction costs, apart from land rent) has the inconvenient consequence of requiring a decentralized labour force, which subtracts from commuter transportation demand. Also housing can hardly be assumed to be an ubiquitous industry, since new housing is likely to be produced on the periphery whereas conversion and replacement will tend to be correlated with age of dwellings (and hence will be fairly centrally located operations). Moreover, since housing is a very durable good, its treatment requires a nontrivial capital sector, nonmalleability, and a dynamic stock-adjustment process that coexists uneasily, if at all, with a long-run equilibrium framework. In addition, with a realistic capital sector, capital–land ratios become a variable, and the identity between site size (the reciprocal of residential density) and amount of housing consumption (in terms of dwelling area) has to be dropped.

It is also possible to treat housing demand explicitly rather than as a derived demand. A rare example of this approach is by Borukhov (1973). Because of his assumptions about the housing supply (a Cobb–Douglas production function under constant returns to scale, a competitive land market but spatially invariant nonland input prices), there is a simple link between house prices and land rents:

$$p_r^H = \beta (p_r^L)^\alpha , \qquad (3.1)$$

where
p_r^H is the house price at location r,
p_r^L is the land rent at r,
α is the land exponent, and

$$\beta = \frac{v^{(1-\alpha)}}{\alpha^\alpha (1-\alpha)^{(1-\alpha)}} ,$$

where v is the nonland input price.

Nevertheless households do not bid for land but purchase one unit of housing each. City size could be measured by aggregating the housing stock, as an identity between the number of households and the housing stock is assumed. Land rents do have an impact on densities by affecting the land-use intensity of development, but this is a decision of the housing producer not of the household. In effect, Borukhov's model assumes that housing demand is inelastic because no household can occupy more than one unit (or divisible parts of a unit) of housing.

Of course, if housing demand is treated explicitly, it is easier to expand the dimensions of residential site choice from the simple, some might say naive, reliance on accessibility to the CBD and site size. In particular, neighbourhood quality and environmental externalities can be taken into account. Although these influences can be accommodated within a demand-for-land model, it is more awkward since there is some controversy as to whether these externalities are fully internalized in land values. In either case, the externalities (for example, clean air, pleasant neighbourhoods) should be entered as arguments into the utility function. Where the demand for land cannot stand as a substitute for housing demand, however, is in relation to dwelling characteristics—house type, number of rooms, and quality variables. Introducing them into a comprehensive general equilibrium model of the NUE type would make it very complicated, however, and a more practical approach might be to deal with this problem via a more restricted housing-market model (for the best example—the NBER model—see pages 185–193).

Transportation
The transportation system in NUE models follows from their basic assumptions. The function of the transportation system is to move commuters from the residential ring to the edge of the CBD; it does not

carry commodities. Presumably, workers either consume at work or (more plausibly, if households are larger than one person) they carry goods home from work. The linear-ray assumption means that transportation is ubiquitous. Since mass transit must be laid out on specific routes, it is too inflexible, so the transportation system has to be a network of radial roads. It must also be a very dense network to satisfy the ubiquity criterion. This is, of course, very unrealistic, since radial road networks tend to have relatively few spokes. A much more efficient coverage of the urban area can be obtained by a network of circumferential *plus* radial roads, but only Kraus (1974) has discussed this as a step towards an optimal layout of the transportation system.

The standard method of dealing with transportation demand in NUE models is to equate it with the number of commuters multiplied by an exogenously specified average road-width requirement per commuter. Transportation enters the model, therefore, primarily as a competitor of housing for land in the residential ring, with the land-use allocation to transportation declining with distance between the edge of the CBD and the urban periphery. It is also possible within the model to allow for congestion, usually by making a component of transport costs an exponential function of the ratio of transportation demand to capacity (that is, by adapting the model associated with Vickrey, 1965).

A major problem with this approach is the assumption of no input substitution in transportation (or making the production of transportation a function of land inputs alone, for example, Mills, 1972a). If the urban model included capital, on the other hand, and if input substitution were feasible, the capacity of the transportation system could be expanded by more capital investment without increasing the demand for land. A mass-transit facility is one obvious possibility; a two-tier freeway is another. In the former case it may be necessary to combine using the transit facility with driving to it, giving rise to multiple-mode work trips that have not been analyzed theoretically. Also, since the transit facility would lie along certain routes, this will distort the spatial distribution of land values, rents, and densities (Capozza, 1973; Davies, 1974), a feature that cannot be analyzed within the framework of the one-dimensional city.

Homogeneity
The typical NUE city is very monotonous. It is located on a flat plain, land is the same quality everywhere, its CBD is simply a factory or work-place owned by a monopolist, any one sector of the city is the same as any other, every house in the residential ring is the same apart from the fact that plot size becomes a little larger with increasing distance, and so on. However, one of the most critical of the homogeneity assumptions refers to the city residents. In most of the early models households were assumed to be homogeneous in incomes and tastes. In later models,

households were allowed to differ in incomes but still had the same utility function. Usually two income groups are assumed. This modification permits the standard prediction that the rich live further away from the CBD than do the poor. The housing-market models used in planning (for example, Herbert and Stevens, 1960; Ingram *et al.*, 1972; Wilson, 1974) tend to disaggregate households into many household types, acknowledging that household preferences vary not only with income but with age of the household head, family size, social status, and other variables. Of course, a theoretical model cannot be expected to replicate the richness and variety of the real world, but some heterogeneity of preferences within an income group might enable such a model to generate the 'split-location' pattern common in many large cities—the distribution where the rich live both close-in, in expensive apartments near to the CBD, and in the outer suburbs at low densities.

The public sector and externalities
Two characteristics of urban economies, that are so distinctive that they need to be kept at the forefront of analysis, are the pervasiveness of externalities and the major role of the public sector. Externalities are associated with the effects of crowding, the benefits of agglomeration (both for production and consumption), and with the provision of public goods, which spills over into the public sector assumptions. The early NUE models tended to ignore externalities but, more recently, they have given considerable attention to traffic congestion (for example, Solow, 1972; 1973a; Oron *et al.*, 1973) with fewer analyses of other externality problems (for example, Polinsky and Shavell, 1973; Fisch, 1974b, on air pollution; Mirrlees, 1972, on residential neighbourhood densities). The general line of the analysis is that these externalities are elements affecting household utility and must enter into locational equilibrium conditions. Also, the equilibrium that results may not be optimal in the sense that changes in the institutional framework, typically the imposition of an externality tax (or subsidy), will lead to a higher maximized utility level.

If road-congestion or air-pollution taxes are to be imposed, then there must be a responsible and statutory authority to impose and collect the tax. Thus a public-sector body, in the form of a transportation planning authority or an air-pollution control board, is implicitly or explicitly assumed. In addition, one or two of the more developed models have adopted the Tiebout (1956) hypothesis at the intrametropolitan level, where the level of public-service provision varies within the city and in some cases (if there are submetropolitan tax jurisdictions) the tax rate also differs, and the associated tax−service mix affects residential choice and locational equilibrium. This type of model recognizes, of course, the existence of a public sector, but its role is treated very passively. It is there because there are public services and taxes, but it makes no attempt to act effectively to influence city welfare. Furthermore there is at least one study (Ohls et al., 1974) that has examined planning controls in the sense of evaluating the impact of zoning on land values.

Despite these recent advances, the public sector remains subdued in NUE models, whereas it is arguable that its role in moulding the urban spatial structure and in urban development is at least as important as that of the private sector. The significance of its functions (in the sphere of zoning, height and other land-use controls, planning permits, urban renewal and housing development, provision of urban infrastructure, planning of the transportation system, public-service provision, tax-levying powers, and as a major land user itself) hardly needs spelling out. One key reason why the public sector is so incompletely specified, of course, is that its objectives and behaviour are difficult to translate into the language of neoclassical competitive-equilibrium models.

Competition in the land market
The underlying assumption of NUE models is that behaviour in the land market can be explained as a competitive process. The land-rent function that results from this process is the allocation mechanism that determines the urban spatial structure as a whole. The equilibrium condition for the critical land-use allocation between transportation and housing is that the land rent at any location should be the same in both uses. Competition between urban uses and agriculture sets the urban boundary condition (and hence determines city size) in many models.

Yet, upon closer examination, the nature of land-use competition in NUE models is strikingly odd. The major form of competition is between housing and roads in the residential ring. The competition between residential and other nonresidential uses is ruled out by the 'exclusive zoning' assumption. This is the reverse of what is important in land-market competition in the real world. Road development is a function of the public sector, with land frequently acquired by compulsion at prices that are lower than competitive market prices. The competitive bidding power of nonresidential activities (especially in the commercial sector) is a major element in land-use conversion from residential to other uses, and is certainly critical to an explanation of the subcentring phenomenon. The exogenous determination of the location of all nonresidential activities—apart from transportation—within the CBD destroys the implicit claim of NUE theories to provide competitive models of the urban land market.

The constraints on land-use competition that are so important to an understanding of how the urban land market operates are ignored in NUE models. These constraints are of several kinds. First, price adjustments may be very sluggish, yet at certain times may take place very rapidly indeed. This arises because prevailing market prices are determined by the supply and demand at a specific time, and these may be wildly out of balance, especially because the supply of available sites is only a very small proportion of the total urban land. Inelasticities in this small volume of land on the market at any one time imply that quite small variations in demand may have a drastic impact on prices or rents. On the other hand, the implicit prices of sites not on the market may adjust only sluggishly. History may have its say, since past prices may have a marked influence on current prices in certain areas. Many sites may be rented on long leases that respond slowly to changes in market conditions. Locational inertia may be strong, making occupiers reluctant to relocate even if transfer to a new site would increase utility or profit. Of course, many of these imperfections are lag and adjustment factors, ruled out by the long-run equilibrium assumptions of NUE models.

Second, the competitive process appears more efficient than it is because of the models' homogeneity assumptions. Although the location of nonresidential activities is not determined competitively, one argument might be that exogenous specification amounts to the same as a competitive outcome, because nonresidential users have both a greater desire for accessibility and the capacity to outbid households, and hence would occupy the CBD in any event. However, nonresidential activities are not homogeneous. There are different types of establishment with very different locational requirements. These include: headquarter and corporate finance and other establishments; large-scale retail firms, wholesalers, and warehouses; small-scale retail stores; professional and personal-service establishments; public facilities; large manufacturing plants; and small manufacturing establishments and workshops. Only a minority of these nonresidential activity types depends heavily on agglomeration economies, and hence requires a CBD location. Since the need for agglomeration is probably declining over time, the single CBD workplace models become progressively less relevant to an understanding of modern cities.

The same kind of argument holds for the residential segment of the land market too. Households are heterogeneous, for reasons other than income differentials, in ways that affect their residential site choice. The importance of race as an influence on this choice, and hence on the spatial distribution of rents, is an obvious case in point. The implications for the smoothness and shape of the rent gradient (Rose-Ackerman, 1975) are very serious. Similarly, heterogeneity of neighbourhoods and differences in housing type (that affect *land* prices since neighbourhood dwelling type has an externality effect on nearby empty sites) imply an only slightly overlapping complex of residential land and housing submarkets, rather than a single metropolitan market. The result is that, even in long-run equilibrium, rent gradients may be discontinuous, and marked variations in residential densities may be found in the same neighbourhood.

Finally, another reason for rent discontinuities and density variations is the role of the planning authority and the effects of planning instruments such as zoning, lot-size standards, height restrictions, building codes and standards, and as a determinant of public facility locations. How important it is to take these into account and, in what ways, depends upon the analyst's assessment of the relative strength of private and public agencies in the urban land market. It is possible to deal with the planning function by introducing zoning and other controls in the form of constraints within the framework of a competitive model, but some might argue that it is necessary to assign a more influential role to the planning function, and if so its operations may be simulated much more easily with some analytical model other than one of those based on neoclassical competitive equilibrium.

Long-run equilibrium
A characteristic of NUE models is that they are static models that focus on long-run equilibrium. The determination of equilibrium is a one-stage allocation problem, and the only way in which time has been introduced is by varying the main parameters in a comparative statics context. There have been one or two deviations from this tradition very recently (see pages 124-129), but not enough to make much of an impression in terms of development of a dynamic model. Dynamic models would need to take account of the durability of urban infrastructure, for example housing, and the fact that current decisions in the land market are in part based upon expectations about future changes in rents. It is unclear whether a long-run equilibrium is attainable, and analysis of the adjustment path towards, but never reaching, an equilibrium might be a more relevant approach.

There is also a controversy, dating back to a well-known thesis by Koopmans and Beckmann (1957), as to whether a general spatial equilibrium is sustainable on theoretical grounds (for a recent viewpoint see Hartwick, 1974). Indivisibilities present a severe problem. An equilibrium assignment can be obtained if the scale of activities is fixed and if there is no locational interdependence. However, neither of these conditions can be expected to hold in normal circumstances. If there is locational interdependence and land is used as an input in production in indivisible amounts, a competitive land market may result in inefficient land patterns. However, the Arrow-Debreu general equilibrium model requires a fixed set of marketplaces where all mobile commodities are exchanged. This condition is usually violated in an urban economy, except under completely static circumstances. If a new locator is added, his location decision may alter the whole set of general equilibrium prices. Also it is doubtful whether he can make a rational decision, since he cannot forecast the effect of his decision on the general equilibrium price levels at different locations. Uncertainties and the possibility of varying attitudes to risk mean that the locational choice is indeterminate (Artle and Varaiya, 1975). For instance, consider a household making a residential site choice. The decisionmaker has imperfect knowledge, his search process is subject to constraints of time and space, and he has no idea about the repercussions of his decision on other movers and relocations. As a result, a multiperiod adjustment process may provide a superior simulation of residential locational behaviour. However, no NUE model has addressed these issues.

These arguments are not intended to denigrate the advantage of the long-run equilibrium approach. It can provide a useful benchmark for the evaluation of actual allocations. Also, a comparison of the long-run competitive equilibrium (the positive solution) with the optimal equilibrium (the normative case), obtained by maximizing some social-welfare function, may shed light on the need for city-wide redistribution through taxes and subsidies to generate a stable optimal solution. NUE models abound with these competitive and optimal comparisons. To the extent that NUE has a normative content, these comparisons have produced some of its more interesting results. Mills and MacKinnon (1973) contrast the impact of these analyses on the understanding of market failure and on prescriptions to correct resource misallocation with the neglect by urban planning of welfare economics.

Mathematical tools
NUE modellers, with a few rare exceptions (Mills, 1972b; Hartwick and Hartwick, 1974; MacKinnon, 1974; Amson, 1974; 1976), have been exclusively concerned with continuous models, retaining the one-dimensional city and smooth-rent and density-gradient assumptions—almost regardless of how uneasily they fit the available empirical evidence. The obstinacy of this faith is not due to an objective evaluation that these assumptions are reasonable simplifications and that continuous models are the most appropriate. Rather, the dominant reason is that it permits the use of familiar mathematical tools, and that because smooth differentiable functions proved useful in analyzing the time dimension it follows that smooth functions should be used in dealing with space. On the contrary, the space economy is characterized by marked discontinuities, and this suggests that in urban economics discrete models—allowing zonal analysis in two-dimensional space (or better still in three dimensions since the skyline of cities can change dramatically in response to variations in capital–land ratios)—may be more profitable. Programming models (linear and nonlinear), fixed-point algorithms, computer simulation, and some interesting adaptations of mathematics used in applications in the natural sciences (Amson, 1974; 1976) are just some of many possibilities.

Yet NUE modellers have doggedly persisted in using the differential calculus with which they have long been familiar. The normative models are usually solved via the use of variational methods (calculus of variations, specifically Pontryagin's principle) or optimal control theory. Each paper goes through the same routine of specifying a utility function, maximizing that function subject to constraints, deriving the first-order maximization conditions using Lagrangian multipliers, and then proceeding from there to solve the model. Although in many cases this approach yields interesting qualitative results, obstacles to an efficient solution are often quite common.

Simple functional forms have to be chosen (for example, the logarithmic utility function, the Cobb-Douglas production function), it is frequently necessary to resort to numerical analysis even to obtain fairly limited solutions, and the models tend to be inflexible with slightly different functions, thus requiring development of the model from scratch again. Even more serious is that the implicit assumptions about the nature of urban phenomena necessary to justify the use of continuous models, and the mathematical methods available to handle these, have never been scrutinized by the analysts. It is this neglect which lends substance to the charge that the conception of cities has been moulded to suit the favoured mathematical tools.

An example of a simple NUE model

To illustrate what the implications of NUE assumptions are for an urban model, it may be illuminating at this stage to describe briefly a simple case[9]. Assume a circular city containing a CBD of radius r_0 surrounded by a residential ring of houses and roads. The outer boundary of the city, $\bar{r}$, could be determined in one of three obvious ways: exogenously assumed; obtained endogenously when all income y is absorbed in transport costs, $t_{\bar{r}}$; or determined by an opportunity cost of land use in agriculture constraint so that boundary rent, $p_{\bar{r}}$, equals agricultural rent, p^a. The city's residents total N families, each with one worker who works in the CBD. The families are identical, with the same income, y, and the same tastes. They attempt to maximize their utility by choosing composite consumption (c, at a numeraire price of 1), amount of housing space, s, and location (measured by distance from the CBD, r) subject to the budget constraint. By assuming a logarithmic utility function, this may be written as

$$\max U(c, s) = \kappa \log s + (1-\kappa) \log c , \qquad (3.2)$$

$$\text{subject to } y = c + p_r s_r + t_r , \qquad (3.3)$$

where
p_r is the rent at distance r, and
t_r is the transport cost for a household living at r.

In the simplest model, transport costs may be assumed exogenous, but in more complex analyses it is common to make these costs endogenous by introducing a congestion effect that makes t_r dependent upon the number of commuters living beyond r. The logarithmic utility function is, of course, a CES (constant elasticity of substitution) function, according to which, no matter where the household lives, a constant fraction, κ, of income net of transport costs is spent on rent, and the remainder on composite consumption. It should also be noted in this model that

---

[9] The example here is based upon Solow (1972; 1973a). Other examples are described in the course of the analysis in various parts of the book.

The standard NUE model

distance is not an argument of the utility function, even though choosing a location is part of the utility-maximizing decision. It does, however, enter indirectly as transport costs in the budget constraint.

The allocation problem is to find the equilibrium rent function, p_r, that maximizes total utility. From the set of assumptions used in this model, in equilibrium all households will have identical utility. It follows, of course, that rents must decline with distance to compensate for higher transport costs. The lower rents further away from the CBD will induce households to substitute cheaper land for composite consumption (the price of which is assumed constant). This means that s_r will increase with distance; therefore, its reciprocal, s_r^{-1}—residential density—will fall with distance.

The first-order maximization conditions are

$$\frac{\kappa}{p_r s_r} = \frac{(1-\kappa)}{c} , \qquad (3.4)$$

and

$$p'_r s_r + t'_r = 0 . \qquad (3.5)$$

Equation (3.5) is the locational-equilibrium condition. For no household to be able to increase its utility by moving, the declining rent gradient must exactly offset the increasing travel-cost function. From equations (3.3) and (3.4),

$$p_r s_r = \kappa(y - t_r) . \qquad (3.6)$$

Substituting equation (3.5) into equation (3.6) yields a differential equation

$$\frac{p'_r}{p_r} = \frac{1}{\kappa}\left(\frac{-t'_r}{y - t_r}\right) = \frac{1}{\kappa}\frac{\mathrm{d}}{\mathrm{d}r} \log(y - t_r) , \qquad (3.7)$$

and this is satisfied by the equilibrium-rent function

$$p_r = p_o \left(\frac{y - t_r}{y - t_o}\right)^{1/\kappa} , \qquad (3.8)$$

where
p_o is the rent at the edge of the CBD, and
t_o are the travel costs at the edge of the CBD.
Since $t_o = 0$ because a resident located at the edge of the CBD does not have to commute,

$$p_r = p_o \left(1 - \frac{t_r}{y}\right)^{1/\kappa} = p_o w_r^{1/\kappa} , \qquad (3.9)$$

where $w_r \; [= 1 - (t_r/y)]$ is the fraction of income remaining after commuting costs from distance r have been subtracted. Similarly space consumption is determined from

$$s_r = \frac{\kappa y}{p_o} w_r^{1 - 1/\kappa} . \qquad (3.10)$$

c

In this model, p_o serves as a constant of integration. Its value can be determined from the given total number of households, N, in the following way. Total housing area within any very narrow annular ring of width dr is approximately $2\pi r b_r dr$, where b_r is the fraction of land at distance r used for housing (the remainder is used for transportation). Assuming that $n_r dr$ represents the number of people living in that ring, in equilibrium the demand for and supply of land must be equal, thus

$$s_r n_r dr = 2\pi r b_r dr . \tag{3.11}$$

Substituting equation (3.10) into equation (3.11) and rearranging, we obtain

$$n_r = \frac{2\pi p_o}{\kappa y} r b_r w_r^{1/\kappa - 1} . \tag{3.12}$$

Since the total population is $N = \int_{r_o}^{\bar{r}} n_r dr$, then

$$N = \frac{2\pi p_o}{\kappa y} \int_{r_o}^{\bar{r}} r b_r w_r^{1/\kappa - 1} dr . \tag{3.13}$$

This expression enables the constant of integration p_o to be determined.

These latter manipulations are helpful in extending the model to the case of endogenous transport costs. If we assume that t_r, rather than being determined exogenously, is made up of a distance cost, t_o, and a congestion cost, this second element can be assumed proportional to traffic density (that is, the ratio of commuters to road width). Thus

$$t_r = \int_{r_o}^{r} \left[t_o + a \frac{N_r}{2\pi r(1 - b_r)} \right] dr . \tag{3.14}$$

Differentiating equation (3.14) and substituting $t_r' = -y w_r'$ gives

$$w_r' = -\frac{t_o}{y} - \frac{a}{y} \frac{N_r}{2\pi r(1 - b_r)} . \tag{3.15}$$

Also since $N_r' = -n_r$ by definition, then from equation (3.12)

$$N_r' = -\frac{2\pi p_o}{\kappa y} r b_r w_r^{1/\kappa - 1} . \tag{3.16}$$

Equations (3.15) and (3.16) are first-order differential equations—the first shows how commuting costs depend upon population density, and the second shows that equilibrium population density depends upon commuting costs (since commuting costs affect the value of w_r). In these equations t_o, κ, a, and y are known constants, and b_r is specified exogenously (in more comprehensive land-use allocation models b_r can be made endogenous). To determine p_o in this case, the following boundary conditions are needed: $N_o = N$; $N_{\bar{r}} = 0$; and $w_o = 1$ (since $t_o = 0$). The model can then be fully solved.

Numerical solutions: the contribution of Mills
Mills has made several notable contributions to the development of NUE. His early paper (Mills, 1967) identified many of the problems that were later to attract the attention of NUE theorists, plus others that have been relatively little explored, such as capital–land substitution. He was the first to examine the question of land-use allocation within the CBD (Mills and de Ferranti, 1971), and his linear-programming model (Mills, 1972b) demonstrated the potential of discrete, as opposed to continuous, models. His monograph of the same year (Mills, 1972a) is the first book-length treatment of NUE, though many NUE-related themes are to be found in Muth (1969) and in earlier writings. Although *Studies in the Structure of the Urban Economy* discusses many issues such as the rationale for the existence of cities, urbanization trends, density gradients, a fixed-coefficients urban model, and surveys previous urban theories, its core is the development of two closely related models of urban structure and their numerical solutions. The first model includes the possibility of congestion (that is, the demand for and supply of transportation services may be unequal), whereas the second model assumes an efficient transportation system (demand equals supply, and transportation services are priced at marginal cost). Although this type of distinction is common to other NUE models, a unique feature of Mills's study is that he tests the implications of the two models with a more thorough numerical analysis. Apart from the other interesting aspects of the research, this feature alone suggests the value of a brief review of Mills's findings.

The city in the Mills's models has more or less the usual characteristics. It is a monocentric city with a CBD, the size of which is determined exogenously (assumed to have a radius r_1). The city's labour force, N, is exogenous and employed only in the CBD. What happens inside the CBD is no concern of the model so that it is, in effect, a model of the suburbs. Outside the CBD, land is used in all directions (apart for a pie slice of $2\pi - \theta$ radians to reflect topographical constraints) for workers' homes and transportation (commuting), so that land rent and population density are functions only of distance. The boundary of the city is found where urban rent, p_r, is equal to the opportunity cost of the use of land, the agricultural rent, p^a.

Housing services are produced with capital and land. The rental rate on capital, p^K, is exogenous and constant throughout the urban area. The determination of the land-rent function is the main objective of the model. It is assumed that housing is produced with a Cobb–Douglas production function under constant returns to scale, thus

$$H_r = A_2(L_r^H)^{\alpha_2}(K_r^H)^{1-\alpha_2}, \tag{3.17}$$

where
H_r is the supply of housing,
L_r^H and K_r^H are land and capital inputs, all at distance r,

A_2 is a constant, and
α_2 is the land exponent.
Input markets are competitive, so equilibrium requires that the value of the marginal product equals the input price, thus

$$\frac{\alpha_2 p_r^H H_r}{L_r^H} = p_r ,\qquad(3.18)$$

and

$$\frac{(1-\alpha_2)p_r^H H_r}{K_r^H} = p^K ,\qquad(3.19)$$

where p_r^H is the price of housing.

The demand for housing per family living r miles from the centre is

$$s_r = B_2(p_r^H)^{e_2} y^{e_1} ,\qquad(3.20)$$

where
s_r is the housing consumption per worker at r,
B_2 is a constant,
y is income per worker, and
e_1 and e_2 are the income and price elasticities of demand for housing.

Income is exogenous, and commuting costs do not enter the housing-demand function (unlike in many NUE models). Total housing demand at each location is simply $s_r N_r$, and equilibrium requires that the supply of housing equals demand at each location, so that

$$H_r = s_r N_r .\qquad(3.21)$$

The model includes the usual locational equilibrium condition—that any change in housing costs with a change in location will be exactly offset by the change in commuting costs, hence

$$s_r p_r^{H'} + p_r^t = 0 ,\qquad(3.22)$$

where $p_r^{H'}$ is the derivative of p_r^H with respect to r, and p_r^t is twice the commuting cost per mile at r (that is, assuming a return work trip).

The transportation system is treated very simply indeed as a function of land inputs alone, so that

$$S_r^T = A_3 L_r^T ,\qquad(3.23)$$

where S_r^T is the road capacity available at r, and L_r^T is the amount of land used in producing these roads. In effect, this amounts to a special Cobb–Douglas production function with a zero capital exponent. Transportation demand is based on work trips alone, and is determined by the location of homes. Thus transportation demand at r equals the number of workers who live beyond r, hence

$$D_r^T = N - \int_{r_1}^{r} N(r)\,dr ,\qquad(3.24)$$

where D_r^T is transportation demand at distance r. Since the use of the urban transportation system by commuters may exceed its design capacity, travel costs should include an element representing congestion cost. Mills uses the Vickrey (1965) formula

$$p_r^t = \bar{t} + \rho_1 \left(\frac{D_r^T}{S_r^T}\right)^{\rho_2} . \qquad (3.25)$$

The first term is the travel cost (twice the sum of operating and time cost per commuter-mile) in the absence of congestion; the second term represents congestion cost by assuming that congestion cost per commuter at r is proportional to a power of the ratio of transport demand to capacity.

All land within the urban area outside the CBD is used either for housing or roads:

$$L_r^H + L_r^T = \theta r . \qquad (3.26)$$

Since land can be used for urban purposes only if it is bid away from agriculture, the urban boundary will be $\bar{r}$ miles away from the city centre where

$$p_{\bar{r}} = p^a . \qquad (3.27)$$

Finally, all workers employed in the CBD have to be housed somewhere in the city, so that

$$\int_{r_1}^{\bar{r}} N(r) dr = N . \qquad (3.28)$$

The solution of the model (for p_r) may be derived as

$$F p_r^\alpha p_r' + \bar{t} + \rho_1 \left\{ \frac{1}{A_3 L_r^T} \left[N - \int_{r_1}^{r} G^{-1}(\theta r - L_r^T) p_r^{-\alpha} dr \right] \right\}^{\rho_2} = 0 , \qquad (3.29)$$

where

$\alpha = \alpha_2(1+e_2) - 1$,

$F = B_2 C^{1+e_2} y^{e_1} \alpha_2$,

$G = \frac{1}{A_2} \left(\frac{1-\alpha_2}{\alpha_2 \beta}\right)^{-(1-\alpha_2)} B_2 C^{e_2} y^{e_1}$,

and

$C = [A_2 \alpha_2^{\alpha_2} (1-\alpha_2)^{1-\alpha_2}]^{-1} (p^K)^{1-\alpha_2}$.

In the second model, transportation services are treated as a competitive industry. The price of transportation at each r is the marginal (equals the average) cost of land resources used, and enough land is devoted to roads to equate the supply of and demand for transportation. As a result, equation (3.25) is eliminated, and is replaced by

$$p_r^t = \frac{p_r}{A_3}, \tag{3.30}$$

and

$$D_r^T = S_r^T. \tag{3.31}$$

The solution in this case is

$$\overline{F} p_r^{\alpha-1} p_r' + 1 = 0, \tag{3.32}$$

where $\overline{F} = A_3 F$.

If we use the condition $p_{\bar{r}} = p^a$, equation (3.32) has the solution

$$p_r = [(p^a)^\alpha + \alpha \overline{F}^{-1}(\bar{r} - r)]^{1/\alpha}, \qquad \alpha \neq -1, \tag{3.33}$$

or

$$p_r = p^a \exp[\overline{F}^{-1}(\bar{r} - r)], \qquad \alpha = -1. \tag{3.34}$$

The remaining problem is finding the value for $\bar{r}$. This is obtained from

$$(1+\alpha)N + \theta A_3(\bar{r} - r_1) - \theta(p^a)^\alpha A_3[(1+\alpha)(p^a)^{-\alpha}\bar{r} + A_3 G]$$
$$\times \{1 - (p^a)^{-1}[(p^a)^\alpha + \alpha G^{-1}A_3^{-1}(\bar{r} - r_1)]^{-1/\alpha}\} = 0. \tag{3.35}$$

The basic parameter values were obtained either from the best empirical estimates or were determined *a priori* on plausibility and internal consistency grounds. The test city was an abstract city of about one million persons, with $N = 300000$. Spatial units are measured in miles, and monetary units in dollars per day. The selected values were: $\alpha_2 = 0 \cdot 20$; $p^K = 0 \cdot 0005$; $y = \$25$; $e_1 = 1 \cdot 5$; $e_2 = -1 \cdot 5$ [not the more common $-1 \cdot 0$ because this resulted in unstable behaviour for equation (3.29)]; $p^a = \$800$ (per square mile per workday); $\theta = 6 \cdot 28$ (that is, a circular city is assumed); $r_1 = 1$; $A_2 = 0 \cdot 01$ (implies a population density of about 3500 per square mile); $B_2 = 0 \cdot 10$ (implies housing expenditure of fifteen percent of income); $L^T = 6 \cdot 25$ (implying twenty percent of the land is used for transportation in a city of ten-mile radius); $A_3 = 40000$ (thus the design capacity of the transportation system is 250000, that is, $6 \cdot 25 \times 40000$, which means a ratio of commuting demand to design capacity of $1 \cdot 2$); $\bar{t} = \$0 \cdot 40$ (the result of an operating cost of $\$0 \cdot 10$ per vehicle mile, uncongested travel speed of 25 mph, and valuing travel time at the hourly wage rate of $\$2 \cdot 50$); $\rho_1 = 1$; $\rho_2 = 2$ (derived from Vickrey). These last two parameters mean a commuting cost per mile of $\$1 \cdot 84$ at the edge of the CBD, which may be rather high at $4 \cdot 6$ times the uncongested commuting cost. The same parameters are retained in the competitive transport model, apart from the travel-cost function. The value of travel costs is endogenous, and obtained by solving equation (3.30); also, in the comparison between the two models $\bar{t}$ is set at zero (assuming that all travel costs are due to congestion). The sensitivity analysis of the experiments focusses on several characteristics that might be compared with available data for real-world cities: urban-area radius; area of the city; average gross population density; percentage of land area outside the CBD devoted to transportation; land rent at the edge of the CBD and three miles further out; and transport cost per mile at the edge of the CBD and again three miles beyond.

Land rent in the model turns out to be about eighty-five times greater at the edge of the CBD than at the urban boundary. Both rent and commuting costs fall off rapidly with distance, though rent declines more sharply. The rent function declines faster than implied by an exponential function, a consequence of the assumptions of very elastic housing demand and a declining commuting cost–distance function. For the sensitivity analysis, the parameters were varied by ±20% (apart from e_1, varied by ±10%, and L^T which could be reduced but not increased). Major findings include the following: an increase in A_2 increases the size of the urban area, thus technical progress stimulates the demand for land-intensive housing services, and induces decentralization; an increase in α_2 decreases the size of the urban area, since the associated increase in land rent results in economizing the use of land to produce housing; an increase in A_3 (technical progress in transportation) predictably allows workers to live further out and promotes decentralization; an increase in B_2 (that is, an expansion of housing demand at given price and income levels) increases the size of the urban area, reduces density and increases congestion cost, but surprisingly reduces rent at the edge of the CBD; an increase in e_1, the income elasticity of demand, has the same effects, except that city size is very sensitive since a ten percent increase in e_1 leads to a ninety percent increase in total land area; an increase in e_2, on the other hand, reduces city size and increases inner land rents via its effect on housing demand. When p^K is increased, land is substituted for capital, and city size falls while density and inner land rents increase. Higher incomes (an increase in y) boost housing demand, expand the urban area, and reduce residential density. Suburban land rents increase, but rent at the edge of the CBD rather surprisingly falls. Since the model is more sensitive to income than to a decline in uncongested travel costs (a fall in $\bar{t}$), it suggests that increases in income have the more important influence on the flattening of the density gradient. The model is in fact very insensitive to changes in $\bar{t}$, and to changes in p^a and ρ_1. What happens with changes in ρ_2 depends on the D_r^T/S_r^T ratio; this explains why an increase in ρ_2 increases congestion cost at the edge of the CBD but reduces that cost further out. A decline in L^T congests the transportation system, reduces city size, and raises land rents. Finally a higher N, as expected, increases both urban area and population density. Of course, it also increases rents and commuting costs; in particular, p_{r_1} is very sensitive to changes in N.

Running the model with $\bar{t} = 0$ (the case for comparison with the competitive transport model) shows that the urban area is larger and has lower densities, land rents, and transportation costs. The comparison with the competitive transport model, however, shows that the latter has a much larger urban area and lower densities (two-thirds more area and one-third lower population density); also, p_{r_t} is much lower. The amount of land devoted to transportation is similar, but the *proportion* is much smaller in the competitive transport model. The reasons are that the model which incorporates congestion uses some land for transportation in the wrong place (in the outer suburbs) and it prices transportation services inefficiently. Thus improper pricing of transport leads to substantial misallocation of resources. Congestion makes commuting costly, and residents try to avoid long commuting by living closer in. This bids up land rents close to the city centre.

Although one of the main contributions of Mills's study is to compare the congested city with the competitive city where transport services are priced at marginal cost (a problem which was being examined about the same time by Hochman and Pines, 1971, and by Solow, 1972, and which has since become an indispensable feature of NUE models), perhaps its most important aspect is its demonstration of the utility of numerical solutions. Many NUE theorists have been content to derive qualitative results. Empirical testing in the real world has been handicapped by the acute shortage of data and by the casting of NUE models in terms of the difficult-to-operationalize utility-maximizing framework. Mills shows that there is another way. Building a model that lends itself to numerical solutions (including sensitivity analysis) provides an effective instrument both for testing theoretical hypotheses and for deriving counterintuitive results. Of course, some of the findings may be due to idiosyncracies of the particular model, but this risk will be minimized if different analysts use numerical methods to solve a wide variety of urban models. Although others have experimented with numerical analysis (for example, Dixit, 1973), and it has since become a less irregular feature of NUE models, it is the detailed research of Mills that really demonstrates its potential. This achievement transcends any quibbles the critic might have with the content and structure of his models.

4

Implications and extensions of the standard model

The treatment of utility
A source of dispute among the new urban economists is how to treat utility among households[10] and with distance from the CBD. Most of the early models started with the assumption of identical incomes and preferences and found it convenient to assume that utility would be equal among city residents. This followed from the assumptions in cases where distance was explicitly omitted from the utility function (for example, Solow, 1972). A departure from this line of analysis was made by Mirrlees (1972) and Riley (1973), with their argument that utility might vary with distance in the optimal town.

The reasons for the argument that optimality may demand inequality are interesting. Locational problems have two peculiar aspects which make them different from other branches of economic analysis. First, there is a concealed nonconvexity in location decisions. An individual can only live at one distance from the CBD. If he is assigned two locations, he can live at either but not both. Second, the social value of land varies spatially because of transport costs and, from a utility point of view, the nonneutrality of distance itself[11]. Households have to make an indivisible locational choice, and their residential preferences may vary. It follows that it is not optimal for everyone to have the same consumption bundle. Of course, for an optimal solution everyone must have the same marginal utility of income. But if the consumption bundles of households vary, they do not necessarily have the same total utility levels.

Although Mirrlees (1972) starts out by assuming that utility decreases with distance, he later shows that utility may increase, may be constant, or may decline with distance according to the assumptions of the model employed. For example, utility will be independent of distance in certain special cases: if all households have the same income, and the utility function is defined to ensure uniform utility at the optimum; if the utility function is homogeneous in composite consumption and land; or if

---

[10] The utility assumptions of NUE modellers have varied. Mirrlees (1972) measured total utility as the simple sum of household utilities, whereas Strotz (1965) maximized a weighted sum of utilities (his model assumed heterogeneous tastes). The welfare assumptions of Riley (1973) and Dixit (1973) are more equity oriented. Riley maximizes the *product* of each person's utility. Dixit uses a constant-elasticity function, maximizing the negative sum of individual utilities raised to an exponent. As the value of this exponent approaches negative infinity, the result becomes the same as that of the Rawlsian criterion (Rawls, 1971) of maximizing infinite utility. The result is that utility is equally distributed.

[11] The marginal utility of distance may be positive or negative. For instance, it will be positive if households flee from pollution, congestion, or crime in the central city, or if they seek quiet and more spacious residential neighborhoods. Conversely, it will be negative if people want to live near the 'bright lights'.

the utility function is logarithmic with respect to consumption and linear with respect to distance and site size.

However, Mirrlees regards uniform utility to be exceptional. If preferences are such that for a given rent gradient an increase in income would induce households to relocate inward, then equilibrium demands that utility should fall with distance. If the utility function separates out goods consumption from space consumption and location, and if transport costs are a deduction from goods consumption, then utility is more likely to be an increasing function of distance the higher are transport costs. The utility gradient depends upon the balance between transport costs and the nature of preferences. Upon certain assumptions, including a transport function that declines steeply from the CBD, utility might first increase and then fall with distance.

Mirrlees argues that the shape of the utility-distance function is a less important consideration than the finding that, except in special cases, utilities will not be equal in an optimal solution: "people cannot, when their locations have to be related to a single central point, be treated identically, and there is no reason, from the purely utilitarian standpoint, why different treatment should lead to the same utilities. Thus the technological desirability of geographically concentrated production activity is, in itself, and apart from all considerations of diverse tastes and skills, a reason for advocating some inequality of incomes" (Mirrlees, 1972, page 123).

Casetti (1971) examined the impact on city population, radius, rents, and densities of an exogenous change in the utility level, subject to the notion of an equilibrium level. This is much more relevant to interurban analysis in which migration equalizes utility between cities than to the case of a single, independent city. Nevertheless, Casetti came up with the interesting conclusion that "everything else being equal, the larger the population of the city, the lower the optimal utility level that its households are able to attain" (Casetti, 1971, page 16). Although this conclusion was derived from specific utility and transport-cost functions, Pines (1972) generalized the analysis to show that this finding is not dependent on specific assumptions. However, his analysis conforms to one of Mirrlees's special cases: households are assumed identical in incomes and preferences, and this means that an equilibrium condition is uniform utility everywhere. Given this, the result follows automatically. In the absence of explicit consideration of the production sector (with the possibility of increasing returns to scale leading to higher incomes from in-migration), an increase in population must reduce utility because it raises central rents, thereby reducing the utility of centrally-located residents and hence the utility of everybody, because of the uniform utility requirement.

Casetti (1973) later examined the particulars of a situation in which the equilibrium and the optimal rent functions were identical. Equilibrium was defined by two conditions: first, that no household could make itself better off by relocating; and, second, the supply of residential land equals the demand. Given Casetti's assumption of identical households in income and preferences, the first condition implies spatially invariant utility, whereas the second requires that all the population be accommodated within the city. He showed that if the utility function was of the following kind

$$U = c^a s^b \exp(-nr) , \qquad (4.1)$$

then competitive bidding for residential land would generate an equilibrium rent gradient identical to the optimum rent gradient imposed by a central planner guided by the objective of maximizing social welfare. This is yet another special case in addition to those discussed by Mirrlees.

Stern (1973) has argued that Mirrlees' special cases of equal utility are even more special than Mirrlees suggested. The most important of these cases is the one where the utility function is homogeneous in goods, c, and land, s. The first-order conditions are

$$u_c = \lambda_1 , \qquad (4.2)$$

and

$$u - cu_c - su_s = \lambda_2 , \qquad (4.3)$$

where λ_1 and λ_2 are the Lagrangian multipliers. The homogeneous utility assumptions imply that

$$cu_c + su_s = bu ,$$

hence

$$(1-b)u = \lambda_2 , \qquad (4.4)$$

and u is independent of distance if $b \neq 1$ (if $b = 1$, there may be many solutions, since changing the size of the population makes no difference to total utility). However, this conclusion applies only to this simple model in which one of many modifications could easily disturb the result.

This can be shown in the following way. According to Mirrlees, the first-order condition (4.3) means that "the effect of adding an extra man to the town should be the same—in terms of welfare—regardless of the place in which space is found for him" (Mirrlees, 1972, page 117). This effect will always include $-(cu_c + su_s)$, but if there is any other cost to the remaining population of adding a man at r that depends on r [say $g(r)$], then the homogeneity-implies-equality argument no longer holds. Equation (4.4) becomes

$$(1-b)u - g(r) = \lambda_2 . \qquad (4.5)$$

The special case holds only if $g(r) = 0$.

There may be several explanations of why $g(r) \neq 0$. For example, if there are environmental externalities for which population density is a surrogate, $g = s^{-1} u_{s^{-1}}$ (usually $u_{s^{-1}} < 0$) which will generally depend on r.

Similarly, if there are road-congestion costs the location of the extra man will affect congestion costs at all points nearer to the CBD than this location. In the Mirrlees model, where k is the road-width requirement per traveller at any point, $g = \int_0^r k u_s \, dr$ since u_s is the marginal utility of the displaced land. A third reason for nonzero g might be variable working hours. If the utility function depends only on goods and land but leisure time is fixed, so that the further a man lives out the less work he does, then g will increase with r because the more distant the location of the extra man the greater the loss in output.

Despite his egalitarian utility assumptions (see page 51, footnote 10), Riley(1973) finds that (after making the simplifying assumptions of a linear transport-cost function and a Cobb–Douglas utility function), in the optimal town, utility increases exponentially with distance. Moreover, assuming uniform wages, Riley argues that social welfare would be maximized if nonwage income is distributed as a Gamma density function (hence the name 'Gammaville'). The degree of inequality under optimal conditions is an increasing function of the elasticity of utility with respect to distance.

Other theorists have taken exception to these results. Oron *et al.* (1973) define the optimum as the maximum utility level which can be realized provided that equals are treated equally. Some households may be worse off in the Riley–Mirrlees towns than under this condition. In this sense their approach cannot be regarded as superior in a Pareto sense. Dixit (1973) goes further and argues that "unequal treatment of identical people is ethically unappealing" and "it is difficult to conceive a viable social system which begins with identical people and then introduces any significant degree of inequality among them" (Dixit, 1973, page 640).

Levhari *et al.* (1972), on the other hand, show that, under certain assumptions about expected utility, social-welfare functions that penalize inequality are Pareto inferior. Their assumptions may be approximated by establishing a lottery system under which entrepreneurs buy houses and distribute them by lottery. Households prefer a competitive equilibrium with lottery to the normal competitive equilibrium. Of course, it does not follow that the Pareto-efficient solution is the 'best'. The most appropriate specification of the social-welfare function depends on societal preferences, and ultimately upon ethics. Equity has a claim to be considered alongside efficiency as one of society's goals.

Transportation land use and congestion

A problem which has received considerable attention from NUE theorists is the allocation of land between housing and transportation (typically roads) in the residential district of the monocentric city. Very few of the studies (a notable exception is Livesey, 1973) have dealt with this question within the CBD. Several of the early articles made the proportion of land used for transportation an exogenous variable. More recently, much

Implications and extensions of the standard model 55

attention has been given to making it endogenous. However, in most cases the framework for discussing this point has been too narrow. The assumptions have been that roads are the sole transport system and that road width is proportional to the number of commuters. In a more comprehensive model that allowed for capital, the amount of land used per person in transportation would depend upon the choice of transportation technology, and this would be the capital-land ratio in transportation.

Much of the interest in the housing-transportation land use ratio was sparked off by Mills and de Ferranti (1971). They developed a congestion model which predicted that congestion would fall off linearly with distance and that near to the CBD it would be possible for all land to be used up for transportation. These findings have been heavily criticized, but the Mills and de Ferranti paper is important for being the first to show that market choices do not result in efficient location and travel patterns. Solow and Vickrey (1971) analyzed the case of the linear city and showed that the proportion of land allocated to transportation is a decreasing, concave function of distance from the city centre. Furthermore, the market value of land would be an imperfect guide in making decisions concerning land use and in particular might result in too much allocation of land to transport near the city centre. Solow (1972; 1973a) later explored this allocation problem in a model which allowed for congestion in the sense that transport costs are determined by traffic density as well as distance travelled. He argued that the allocation between housing and transportation should be a 'happy medium', defined by that allocation which minimized total rents (or, in his model, which minimized CBD rent since this determines the rent gradient). Similar results were obtained by Fisch (1974a) and Hochman and Pines (1971). The characteristic of these optimal solutions is that less land is allocated to transportation at close-in locations than would result from market forces. The implication is that a degree of congestion in the central city may be consistent with optimality (see also Mills, 1972a).

Livesey (1973) dealt with the land-use allocation between housing and roads in more detail. Allowing for the existence of a nontrivial CBD, he showed that the transportation land-use allocation would be a monotonically increasing concave function of distance up to the boundary of the CBD, when it becomes a monotonically decreasing convex function and reaches zero at the city boundary. Transportation cannot absorb all land near the CBD boundary, because the obvious solution would be to expand the CBD (ruled out by Mills and de Ferranti because they fixed the size of the city exogenously)[12].

---

[12] Dixit (1973) performed some numerical calculations which showed that no matter how large the city (even double the optimum size) the proportion of the area used for roads never reached unity, though in some cases it reached 0·75 close to the CBD.

Legey *et al.* (1973) also examined the question of land-use allocation between residences and transportation, and between the CBD and the city. In comparison with the other studies, its distinctiveness was that it allowed for capital-land substitution. However, to simplify the analysis, the demand side was trivialized—demand for housing and transportation was assumed completely inelastic. The model's key variables are: size of the CBD, amount of land in housing, amount of land in transportation, capital in land, and capital in transportation. Transportation costs increase with traffic density. The total social cost is the sum of interest on capital, transportation cost, and the opportunity cost of land. Two institutional arrangements are compared. In the first, a central planning authority determines the land-use allocation that minimizes total social cost: this is the optimal allocation. In the second, allocation is where the rent profile is determined by market forces, and landlords allocate capital to housing so as to maximize profits, whereas the city government allocates land and capital to transportation so as to maximize net benefit: this is the market solution.

The market city is more spread out than the optimal city, that is, it has a flatter density gradient. It tends to devote too many resources to housing near the periphery and too many to transportation close to the city centre. If population grows, however, the ratio of the optimal- to the market-city size asymptotically approaches a constant less than unity. The market cannot sustain the optimal allocation, but the imposition of congestion tolls in a market system can, since there are many types of taxation schemes that can convert the profile of the market rent to an optimum. Although the possibilities of capital-land substitution had been explored earlier (Mills, 1967; Hoch, 1969), the innovation of this study was to do this in the presence of externalities. The limitations of the model (admitted by the authors) are: its primitive treatment of demand; the standard, but restrictive, assumption that rent is determined solely by access to the CBD; and its neglect of the durability of capital and the lack of *ex post* substitution (see Anas, 1976b).

Riley (1974) showed how the road width requirement per person—the standard unit for determining transportation-land allocations—is amenable to road technology, with the optimal technology being determined by the equation of its costs with the benefits, measured in terms of reduced travel time. He even considers the implications of staggered working hours for his model. Finally, Kraus (1974) extends the problem of the optimal allocation of land for transportation to a two-dimensional circular city with a network of radial and circumferential roads, which introduces a new variable—a choice of routes. Assuming that travel demand between origins and destinations is inelastic, optimizing the road network's pricing system is equivalent to optimizing route use. He also shows that if the average cost per trip mile is an increasing, strictly convex function of traffic density, there is no travel through the CBD in an optimal city. Indeed, in the absence of an inner ring road, any traffic configuration that induces trips through the city centre leads to explosive travel costs.

The standard recipe for dealing with a divergence between the competitive and optimal allocations of land to transportation and other uses is to impose a congestion toll equal to the gap between the marginal social cost and private cost of land use, and to redistribute the proceeds as a lump-sum subsidy. As shown by Oron *et al.* (1973), among others, this will result in a more centralized city than under competitive conditions (the toll reduces travel demand[13], and induces people to live closer in).

Rather surprisingly, the analysis of congestion tolls and its impact on urban spatial structure antedates NUE by several years, as the first paper to examine this problem within a welfare-maximization framework was by Strotz (1965). This little known paper was a pioneer in the field of NUE, neglected in the middle 1960s—perhaps because the urban economics profession was not yet ready to take advantage of its implications, but it anticipated many of the conclusions of more recent work, especially Mirrlees' results (1972).

Strotz's analysis is based upon the examination of seven cases that he calls 'parables'. He summarizes his conclusions: "a stretch of road should be taxed, subsidized, or be exactly self-financing depending upon whether returns to scale of the congestion function are adverse, favourable, or constant; this rule applies whether we have a single road (first parable), a multiplicity of roads (second parable), a road subject to variation in the flow of traffic (third parable), whether the purpose of travel be pleasure or work (fourth parable), whether residential space be considered explicitly (fifth parable), whether the road itself takes up space (sixth parable), or whether there are external economies to the concentration of productive activity (seventh parable). However, in the last-named case a subsidy should be paid for each work trip that is made, regardless of the distance traveled" (Strotz, 1965, page 169). This last prescription arises because of the simple assumptions made about the nature of external economies in production. He assumes that an individual's earnings depend on his trips to the city centre and on the total number of trips to the CBD made by all

[13] In an interesting paper, Henderson (1974a) argues that in certain circumstances congestion tolls may induce greater efficiency via reducing travel cost, decreasing congestion, and *increasing* the number of road users. This is because the travel decision not only involves choice of mode (and in long-run equilibrium a possible change in location) but also a time at which to travel. Congestion tolls affect relative prices at different travel times, since the cost of travel at different times is a function of road congestion at different times and of the cost of arriving at the destination at the most desired, relative to less desired, times. An earlier study by Vickrey (1969) merely considered the decision regarding travel time in terms of queuing at a bottleneck, and assumed that total trips and costs were fixed.

others, but keeps total city population fixed. However, the case for a subsidy stands if a different set of assumptions are used. For example, if there are continuous scale economies of urban production, and city size is treated as a variable, it would pay to subsidize migration into the city (Lave, 1970; Mirrlees, 1972).

Strotz's conclusions about the congestion tax anticipate the results of several later studies (for example, Oron *et al.*, 1973; Riley, 1974; Solow, 1973a; Legey *et al.*, 1973; Mirrlees, 1972; Dixit, 1973; Kanemoto, 1975). In particular, Strotz discovered the critical result that congestion tolls make the city more compact—a finding replicated many times in the intervening years. Strotz also examines the equilibrium conditions for the use of land for roads. The cost of land must be based on the equilibrium rent as determined by the supply of and demand for land for all uses. This is the problem subsequently examined by Solow and Vickrey (1971), Mills and de Ferranti (1971), Livesey (1973), Sheshinski (1973), and Riley (1974)[14].

Transportation alternatives—mass transit

Most NUE models have assumed a ubiquitous road network as the prototypical urban transportation system. Hartwick and Hartwick (1972) presented a formal analysis of the residential distributional changes that might follow the opening up of a radial arterial road, whereas Davies (1974) undertook an empirical examination of the changes in residential density that followed the opening of the first subway line in Toronto in 1954. This subway was constructed through a low-density corridor, and after some seven years (shown in 1961 Census data) there was a marked increase in residential densities close to the location of the subway (distance from the subway was calculated as the minimum distance between the subway *line*—not stations, although the stops were very close together—and the centre of the Census tract).

For a comprehensive theoretical evaluation, Capozza (1973) is the only major study. He developed a general equilibrium model of the city with two transportation modes—a land-intensive mode (roads) and a land-economizing mode (subways). The two major conclusions were: first, that if a subway is added to an existing road system of a city, it has a suburbanizing effect on the city; second, the development of a subway has a considerable impact on house prices and land values, but their precise form is not easily predictable.

---

[14] Although outside the scope of the present discussion, Strotz's work is innovative in at least two other respects. First, he does not assume identical tastes among households; this may complicate the analysis, but it is a bow to realism ignored in most more recent work. Second, his assumption of increasing returns to scale in production is consistent with the reasons for the existence of cities, but most of the early NUE models persisted in the analysis of the unsatisfactory case of constant returns.

He starts by comparing a road city to a subway city, and identifies four different effects. First, the road city requires more land, especially close to the CBD, and hence is more suburbanized. Second, if the cost to the road user is less than the marginal cost of road use (the status quo), the effective wage is higher in the road city and it will grow relative to the subway city via in-migration. Third, the marginal costs of transportation are constant in the subway city but increase with proximity to the CBD in the road city. This will hold true even if road use is incorrectly priced, owing to the time costs of traffic congestion. The result is that the rent gradient in the road city will be more convex. Last, the proportion of population working in the CBD will be higher in the subway city. A possible reason, cited by Capozza, is that the subway is efficient for carrying passengers, but does not carry freight. Accordingly businesses in the subway city must agglomerate to minimize interfirm transportation. A more satisfactory reason, discussed later by Capozza in another context, is that because of economies of scale in mass transit, marginal costs of transportation may *decline* for subway users but they increase with scale for road users. For a given city size, the subway city will be more centralized in terms of jobs because, to the extent that commuting costs are reflected in wage levels, CBD labour costs will be lower in the subway city than in the road city.

The more relevant model is not so much the subway versus the road city but the hybrid city in which roads and a subway both exist. In particular, what happens when a subway is added to an existing road system is especially critical. The immediate effect, compared with the alternative of widening the existing road or building a new one, is that rents around the subway will be *lower* because the subway uses less land. The longer-term effects, however, will correspond to the typical transport-improvement model that predicts increased suburbanization in the form of flatter gradients for rent and density, and in the number of people living beyond a given distance. This result is contradictory to the popular argument that subways would help to revive the central city, and may appear counterintuitive. However, the problem is more complex. There may be localized increases in densities around the subway as redevelopment takes place (this makes Capozza's theoretical finding consistent with Davies's empirical research mentioned above). But the critical argument is that a subway, like any transportation improvement, reduces costs and time of travel for people living at its terminal and beyond, and hence encourages suburbanization. This is not incompatible with the view that, starting from scratch, a city based on a freeway system will probably be more decentralized than a city based on subways, partly because of its greater land demands, partly because the network may have more arteries and favour dispersal. Also, probably, there are high switching costs involved in transferring modes. Obviously a subway system will only have the predicted effects if it is used, and associated measures (for example, easy parking facilities at subway stations) to minimize switching costs will be needed to justify major investments in mass transit, as well as incentives to switch modes, such as very heavy CBD parking charges.

Two additional points of interest emerge from Capozza's analysis. First, introducing a subway, like almost all major transportation changes, will have redistributional effects. In the short run the users of the facility gain, but in the long run perhaps only the landowners gain since most of the benefits will seep into land rents. Second, the subway is sometimes proposed as a preferable alternative to a freeway because the latter creates such traffic demand that congestion will soon be as bad as ever. To the extent that this argument is valid, it also applies to the subway. The reduced congestion costs associated with a subway encourages more use by the existing population and attracts new population. Moreover, even if capacity expansion is more elastic in the case of the subway, the effects of subway expansion will still include higher house prices. The case for a subway or other mass-transit facility, relative to further road extensions, is a much more complex problem than the vocal antifreeway lobby contends.

Externalities
(a) *Air pollution*

A question of considerable theoretical interest, and one that is having major implications for empirical analysis, is the extent to which environmental quality characteristics, such as clean air, and other externalities are capitalized in land values (or internalized in rents). Many empirical studies have attempted, using multiple-regression techniques, to obtain an estimate of the value of clean air (for example, Ridker and Henning, 1967; Anderson and Crocker, 1971; Wieand, 1973). Freeman (1974) has shown that this is fraught with difficulties. The problems may be illustrated with a simple diagram.

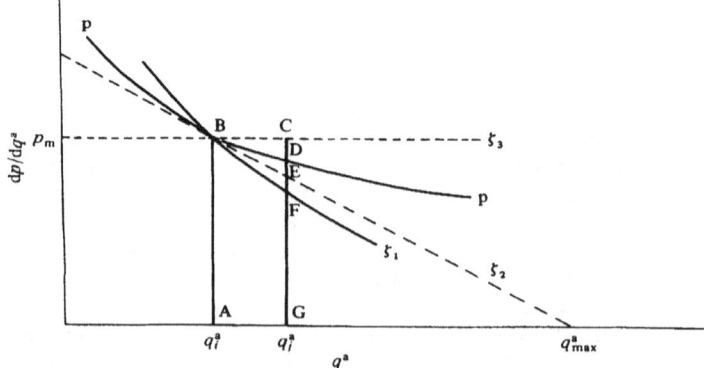

Figure 4.1. Benefits of improvements in air quality.

Figure 4.1 shows the relationship between the marginal costs and the marginal benefits of improvements in air quality (since clean air may be very difficult to measure directly, in empirical studies the problem may be couched in terms of reductions in air pollutants). The curve pp is the marginal-cost function. This is the slope of the rent/air-quality function, $p(q^a)$, where p is the rent and q^a is the air quality, obtained as the partial relationship between rent and air quality in a multivariate analysis of rents, on the assumption that the rental market is in equilibrium. The function ζ_1 is the marginal-benefit function (indicating marginal willingness to pay for clean air) for a household, given real income and holding other prices constant. The intersection of ζ_1 and pp, at B, is the equilibrium utility-maximizing position for the household, where the marginal costs and benefits of clean air are equal.

By using a model of hedonic price indices, it is possible to obtain an estimate of p_m in figure 4.1 [15]. However, this only measures one point on the marginal-benefit (demand) curve. If we consider a nonmarginal improvement in air quality (from q_i^a to q_j^a in figure 4.1), the benefit to the individual household is the area under the marginal-benefit curve ζ_1, that is, area ABFG. But ζ_1 cannot be measured directly. Accordingly the analyst has to make certain assumptions about the shape of the ζ function. The simplest assumption, that marginal benefit is constant, was implicit in Ridker and Henning's study (1967). This means that the ζ function is horizontal (ζ_3 in figure 4.1), so that the benefit of the improvement in air quality to the individual is given by the area ABCG. This assumption probably overestimates the benefits of clean air. An alternative is to assume that the ζ function is downward-sloping but linear (ζ_2), intersecting the horizontal axis at q_{max}^a (equivalent to zero pollution). In this case the benefit is ABEG.

There is one set of circumstances in which the benefit of the improvement in air quality may be directly measurable without arbitrary assumptions. This is if the rent function with respect to air quality can be interpreted as the demand curve for clean air. This assumes that the externality benefits of clean air are fully internalized in the land market. If this is so then, in terms of figure 4.1, the pp and the ζ functions exactly coincide, and the benefit per household of cleaner air ($q_j^a - q_i^a$) is the area under the pp function, ABDG. However, this interpretation requires certain restrictive assumptions. A sufficient condition for this result is that all households have equal incomes and identical (Cobb-Douglas) utility functions (Polinsky and Shavell, 1973). If each level of air quality corresponds to a particular location, the rent function is—given the above condition—that which makes all households indifferent to location.

---

[15] This requires a two-step procedure that makes use of cross-sectional data for many cities, unless one is willing to make the simplifying assumption that all households have identical demand curves for clean air (Freeman, 1974, page 78).

The rent function is then the demand curve for clean air, and clean air is internalized in the land market.

This simple analysis shows that the change in rents or land values associated with a given improvement in the quality of air (or any other externality change) will not usually measure the benefit of the improvement, except under specific, restrictive conditions. The actual benefit—when these conditions do not apply—will be less than the change in rents. The case where the benefit exceeds the change in land values (ζ_3 in figure 4.1) is inconsistent with equilibrium, since it fails to satisfy the second-order condition that the marginal-benefit curve cuts the marginal-cost curve from above.

Using a model very similar to that adopted by Solow (1972), Fisch (1974b) has analyzed the impacts of automobile air pollution on urban structure under three different institutional assumptions: when the urban resident bears the cost; when the motorist bears the cost; and the case of compulsory emission-controls. The specific assumptions are a monocentric city of homogeneous individuals, and the linear, logarithmic, utility function

$$U(c, s) = g \log c + h \log s \,, \qquad (4.6)$$

subject to $y = c + p_r s_r + z_r^l$, $\qquad (4.7)$

where z_r^l is the locational cost at distance r, consisting of two elements—transport costs, t_r, and air pollution costs, z_r^{ap}. It is assumed that every individual commutes to the CBD by car, so that the cost of pollution is a function of the level of traffic, where this is defined as the number of people living beyond a given distance r (that is, $N - N_r$).

Table 4.1 shows the measure of pollution costs, b, in the three cases. In case 1 the pollution costs caused by commuters beyond r fall upon residents between the edge of the CBD and r. In case 2 the pollution costs fall upon the motorist; that is, those living beyond r. In case 3 pollution damage is avoided, but at the expense of a lump-sum cost, z^{ec}, for an emission-control device and higher operating costs ($t^{ec} > t$) due to lower fuel economy when a device is installed. The second row in table 4.1 presents the city sizes under the three assumptions. Since city size is determined exogenously, the urban boundary is located where all income

Table 4.1. Impacts of automobile air pollution.

	Resident pays (case 1)	Motorist pays (case 2)	Emission control (case 3)
Pollution cost	$b_1 N_r$	$b_2(N - N_r)$	$z^{ec} + (t^{ec} - t)r$
City size	y/t	$(y - b_2 N)/t$	$(y - z^{ec})/t^{ec}$
Maximum utility	$(g+h)\log(y - b_1 N)$ $-h \log p_1$	$(g+h)\log y$ $-h \log p_2$	$(g+h)\log(y - z^{ec})$ $-h \log p_3$

net of pollution costs is absorbed by transport costs. The third row shows the maximum utility for residents located at the edge of the CBD (and since locational equilibrium is assumed to require constant utility everywhere), and each entry defines the utility for everyone in each situation respectively. Utility differs for two reasons: the relevant net income varies, and so does the land rent at the edge of the CBD.

Fisch evaluates the impacts of alternative institutional arrangements under three headings: population density and city size; political fragmentation; and the distributional effects of pollution abatement. The present arrangement (case 1) results in a larger city and in a lower population density than either of the other two possibilities. If the residential area is a single jurisdiction, utility is greater under case 1 than under case 2. This is because lower net income (at the edge of the CBD) is more than offset by lower land rent. However, if the residential district of the city is divided into two jurisdictions, the inner-city group would prefer case 2 whereas the outer ring would prefer case 1. This is a good example of the conflict of interests between the central city and the suburbs. The residents in the inner city obviously prefer case 2 because this shifts the burden of air pollution on to suburban commuters. If the cost of the emission-control device is a policy variable, adjustable by the government and not constrained by its production cost, then a comparison of the utility of case 1 with that of case 3 shows that there is a lump sum, z^{ec}, that will confer equal utility. The interesting question, however, is what happens when the assumptions of fixed population and homogeneous incomes are relaxed. For a given control cost, a larger city will prefer case 3 whereas the smaller city prefers case 1. Similarly, the higher-income city will prefer case 3. The approach by way of an emission-control device is regressive with respect to both population size and income distribution: in other words, the larger and/or the richer the city the greater the payoff from this solution. Apart from this being an outcome of applying the model, the result is predictable since automobile pollution is worse in big cities, whereas the demand for clean air is income elastic.

There have been a few other studies of the impact of air pollution on urban spatial structure. Oron et al. (1974) examine the effects of pollution and other nuisances generated by motor vehicles. The major finding, apart from familiar results similar to those obtained with other externalities, is that efficient pricing of the mobile source of pollution limits suburbanization, whereas efficient pricing of a fixed source at the city centre encourages suburbanization (Oron and Pines, 1975). A higher pollution rate induces a larger market city but not a larger optimum city. This is consistent with Fisch's analysis.

Strotz and Wright (1975) have analyzed the case of industrial pollution by using a model in which the utility function contains pollution and the supply of labour (both negatively) and general area preferences, in addition to composite consumption and space. Individuals have the choice of living further away from the pollution source and incurring higher transport costs. In a multiple-factory city, a pollution tax will be needed to allocate resources efficiently and to allow firms to maximize profits in a manner consistent with the maximization of social welfare. On the other hand, in a one-factory town with a nationally mobile population the external diseconomies of industrial pollution will be fully internalized in land rents and wages, and no pollution tax is necessary (this is the small, open-city model; see pages 65, 66).

(b) *Public goods*

In addition to the concept of externality rent (see pages 150-157), it is possible to take a standard NUE model and show how the introduction, in very general terms, of public goods and externalities alters the problem. Intuitively one would expect that variations in the spatial distribution of public services or of disamenities would affect the equilibrium values of urban land rents and/or wages. For instance, a common explanation of higher wage levels in big cities, and their consistency with long-run equilibrium, is that wages have to be higher to induce migrants (or retain residents) in order to compensate for negative net externalities due to pollution, congestion, and higher crime risks (Hoch, 1972; Tolley, 1974). Within cities, areas receiving higher levels of public services would have higher rents and would pay lower wages. Conversely, extensively polluted areas would exhibit lower rents and would pay higher wages. Although intuition is not wrong in these instances, generalization is dangerous. Much depends on the nature of the city; in particular, whether it is small and open or whether the framework assumes a closed city.

A model for dealing with these problems has been suggested by Polinsky (Polinsky and Shavell, 1973; Polinsky and Rubinfeld, 1974). The idea is to convert the standard utility-maximizing model to include public goods and/or other externalities. For example, in the public-goods case, the problem may be stated as

$$\text{maximize } U = U(c_r, s_r, o_r) , \tag{4.8}$$

where o_r are the public services at each location,

$$\text{subject to } y = vc_r + (1+g)p_r s_r + t_r , \tag{4.9}$$

where
g is the property tax rate, and
p_r is the net-of-tax price per unit of housing (land) services at r.

In the case of air pollution, the formulation might be

$$\text{maximize } U = U(c_r, s_r, q_r^a) , \tag{4.10}$$

$$\text{subject to } y = vc_r + p_r s_r + t_r , \tag{4.11}$$

where q_r^a is the level of air quality.

Thus the method involves introducing the externality as an argument of the utility function. In the first case the budget constraint is also affected, but in the second case it remains the same. There are several applications of this modified model. One example, easily illustrated with the second case, is the impact of air pollution on property values (for further discussion of this fascinating problem, see pages 60-62). The procedure involves converting the utility function, which is difficult to operationalize, into an *indirect* utility function, U^i, expressed in terms of prices and income (and in this particular case, air quality). At a given location r, equation (4.11) is solved for the market-demand functions for housing and the composite good, and the demand functions are then substituted into the utility function. Since the indirect utility function is related to each specific location, the price of the composite consumption good does not enter into it since this price is assumed to be the same everywhere. The indirect utility function is

$$U_r^i = U^i(p_r, y - t_r, q_r^a), \tag{4.12}$$

and the individual then chooses the location that maximizes U_r^i. If individuals are free to move among locations, the equilibrium must be such that no one could increase his welfare by moving. Thus, in equilibrium, a utility level, U^{i*}, must exist that is independent of location, hence

$$U^{i*} = U_r^i = U^i(p_r, y - t_r, q_r^a). \tag{4.13}$$

Moreover, implicit in equation (4.13) is an equilibrium relationship between property values (land rent), p_r, on the one hand and U^{i*}, t_r, and q_r^a on the other, thus

$$p_r = f(U^{i*}, y - t_r, q_r^a). \tag{4.14}$$

If individual action alone (as opposed to the behaviour of *all* individuals) cannot affect house prices and travel costs, then t_r and q_r^a can be assumed to be exogenous. For a direct link to be made between property values and air pollution, the equilibrium level of utility level U^{i*} must *not* be influenced by conditions elsewhere in the city.

Under what circumstances will this requirement hold? Briefly, it will hold if the city in question is open and small. If the city is open with perfect mobility at zero cost between cities, a common utility level will prevail throughout the system of cities (both between and within cities). Also, if the city is so small that its development has a negligible impact on the system as a whole, then U^{i*} is exogenous. It is not necessary to solve a general equilibrium model, and p_r is a function solely of locational characteristics at r. An estimated partial relationship between p_r and q_r^a makes sense.

However, if migration is not perfect and costless, or if the city is not small, then improvements or deterioration will affect the equilibrium utility level U^{i*}. From equation (4.14), the change in p_r depends directly on the change in air quality at r, and indirectly on changes in air quality throughout the city via the effect on U^{i*}. The direct effect of a change in q_r^a and the indirect effect of the associated change in U^{i*} are in the opposite direction. The value of p_r is a function of the characteristics of all locations, not merely of location r. To predict the precise impact of a change in q_r^a on p_r it would be necessary to solve a general equilibrium in which both U^{i*} and p_r were determined endogenously.

When the assumption of a small, open city is feasible, it allows a host of important analytical problems to be examined. For instance, Polinsky and Rubinfeld (1974) explore the impacts that follow a change in the property tax and/or the level of public services. For example, if the tax rate is increased and public service levels are held constant, the wage rate increases and the net-of-tax house price falls. Also the city becomes smaller in area, and per capita housing consumption falls. The change in total population, on the other hand, is indeterminate. The burden or benefits of local fiscal changes fall upon owners of the one immobile factor, land. The level of utility is fixed exogenously and hence cannot be affected by intraurban changes. The assumption of the small, open city and the device of the indirect utility function are important simplifications for the purpose of deriving analytical results.

Leisure, family size, and other modifications

Beckmann (1973) has developed some interesting modifications of the standard model. One variant is to introduce leisure time, θ^ℓ, as an argument of the utility function

$$U = U(c, s, \theta^\ell) , \tag{4.15}$$

where transport costs consist of time costs, θr, in which θ is the time per mile (in minutes). Assuming a logarithmic utility function with a_1 as the land exponent and a_2 as the time exponent, the rent function may be derived as

$$p_r = p_o(\Theta - \theta r_o)^{-a_2/a_1}(\Theta - \theta r)^{a_2/a_1} , \tag{4.16}$$

where
Θ is the total available time,
p_o is the rent at the edge of the CBD (also the constant of integration), and
r_o is the CBD radius.

Rents decline to zero where $\bar{r} = \Theta/\theta$, that is, where all available time is used in commuting. This is an alternative boundary condition to the usual equality between urban boundary rent and agricultural rent, or absorption of all income in transport costs. If rent is measured from $\bar{r}$, it rises as a power function of distance. The rent function is steeper the higher the value of the time exponent, a_2, and the smaller the value of the space exponent, a_1. Of course, if a_2 is large and a_1 is small, people cannot easily be compensated for lost time with more space; accessibility is highly prized.

Another extension is to allow—in a simple way—for decentralized employment. If a fraction of the total labour force is employed outside the CBD in service jobs (for example, retail trade) and distributed in proportion to local population density, this is easily accommodated provided that they receive lower wages. A wage gradient must be introduced so that CBD workers are compensated with higher wages for having to commute to the CBD. The number of work trips plus shopping trips can be permitted to vary by introducing a gravity-model effect in which the number of trips is inversely proportional to distance. The number of days worked can also be allowed to vary. The predictable result is that households living far from the CBD prefer to sacrifice consumption (paid out of wage income) for more leisure. The introduction of a continuous income distribution (represented by a Pareto distribution), far from being a recent innovation, was of course discussed in Beckmann's first published NUE paper (Beckmann, 1969). The result is that in an unbounded one-dimensional city with a logarithmic utility function, rent, density, and income are all power functions of distance.

Most interesting of all, Beckmann examines the influence of family size on household location. If there are m commuters and n is family size, and household utility is measured by the sum of utilities of its members, then introducing this change into the model described above yields the demand price for housing, thus

$$p(m/n, r) = p_o(m/n)(\Theta - m/n\theta r)^{a_2/a_1} \qquad (4.17)$$

In this case there are discrete land zones, one for each household type described by m/n. These zones are ordered in terms of decreasing transport cost per family, that is, as m/n falls. Thus single persons and childless couples live close to the CBD, whereas large families with few working members live on the periphery. The result is obtained without introducing the factors usually referred to as explanations of families living in the suburbs—preferences for large gardens, suburban life-styles, and access to open space. Equation (4.17) describes rents within each zone. At the zonal borders, rent is continuous (to satisfy the locational-equilibrium condition). Since discrete social classes or income groups can be analyzed with the same kind of model, Beckmann suggests that this approach is simpler than the mathematical complexities raised by his earlier continuous income-distribution model (Beckmann, 1969).

5
The monocentric city

Agglomeration economies in the CBD
A dominant feature of the NUE spatial-structure models is that they neglect the allocation of land use within the CBD. Even when the original assumption of a point CBD was relaxed to one of finite (and very often fixed) radius, what happened in the CBD was not considered of interest. NUE models are, in effect, suburban models dealing with residential rents and population distribution between the boundary of the CBD and the urban periphery. The exclusive-zone theory makes compartmentalization easy. If no houses are found within the CBD and no workplaces are found outside it, it may be more justifiable to treat the two zones as separable. On the other hand, if nonresidential establishments intermingle with housing outside the CBD, then the CBD and the rest of the city have to be considered together to determine locational-equilibrium conditions for businesses.

Another consequence of downplaying the CBD is trivialization of the conditions of production. In some early models CBD employment was treated as exogenous; in others, the production assumptions were that one commodity is produced under constant returns to scale. These are rather weak assumptions. Perhaps the main reason why cities exist (or at least why they grow) is increasing returns to scale from contiguous production (Koopmans, 1957, page 154; Mills, 1972a, pages 5-9). The size of the city can then be related to the extent of increasing returns (Mirrlees, 1972; Starrett, 1972).

However, if increasing returns are treated solely as a production phenomenon rather than via the broader but more elusive concept of agglomeration economies, then in a one-good model the CBD in equilibrium will consist of a single producer—a monopolist. The first detailed analysis of this situation was by Dixit (1973), who assumed a production function of the type

$$Q = BE^\alpha A^\beta , \qquad (5.1)$$

where
Q is output,
B is a constant,
E is man-hours of labour,
A is the area of the CBD ($A = \pi r_o^2$, where r_o is the radius of the CBD), and $0 < \alpha, \beta \leq 1$, but $\alpha + \beta > 1$.

His model has the strange result that all land within the CBD is used for production so that workers commute to the CBD, and once they cross the border they are in the factory and at work. One way of conceptualizing this is to imagine that workers stand at the CBD boundary and pull levers; another, a little more plausible, is that there is circulation space within the factory and the provision of this is part of producers' costs. Since neither is very convincing, it is more sensible to allow for transportation within the CBD. This has been attempted in several studies (Mills and de Ferranti, 1971; Sheshinski, 1973; Livesey, 1973; 1976; and Alao, 1976).

Despite its limitations, Dixit's model can generate interesting results. First, his model is more general than most of its predecessors. He does not fix the size of the CBD, the city boundary, the urban population, or the transportation land-use requirements in advance, but allows these to be policy variables. Second, the main preoccupation of his model is to determine optimal city size (defined as the city size that maximizes total utility) as the net outcome of increasing returns to production on the one hand, and diseconomies of transportation due to congestion on the other. This question is examined by using numerical examples based on varying the sum of $(\alpha+\beta)$, but keeping the α/β ratio constant at 2·5 (realistically too low, but chosen to avoid running into the problem of increasing returns to labour alone). B is allowed to vary so as to keep Q unchanged.

Three specific cases are considered, where $(\alpha+\beta)$ is equal to 1·00, 1·15, and 1·35 respectively. In the first case (zero economies of scale) utility is maximized when population is zero, so that a completely dispersed population is optimal. In the second case utility is maximized at a city size of about 150000 (of course, the actual numbers depend on the assumptions made about the values of other parameters). In the third case the optimal city size is about 200000. The first increments of scale economies lead to rapid increases in the optimum population, but the effect wears off rapidly. In the Dixit model, 200000 households are about the maximum. "Thus the model cannot explain Chicago as an optimum town. This is probably equally the fault of the model and of Chicago" (Dixit, 1973, page 648).

The model also suggests that ignoring congestion and using another index of welfare, such as wage income, can be very misleading, especially at low levels of scale economies. The radius of the city is a decreasing function of city population, and eventually declines in very big cities, because long commuting under congested conditions would take too much time and subtract from the level of potential output. Although the optimum population is relatively insensitive to changes in rents, it is very sensitive to changes in transport costs. Lower transport costs permit more scale economies, and cities can therefore be much bigger. The optimum road width increased for some distance into the residential ring (usually one-third to a half), then decreased until it reached zero at the city boundary. Even in very congested cities it never absorbed all the land, though in extreme cases it might take three-quarters of it near the CBD boundary.

The two major defects of Dixit's model are fully admitted by him. First, transport costs need to be introduced within the CBD (this is the criticism mentioned above), which would lead to a rent gradient within the city centre, and then it becomes sensible to relax the assumption of a fixed land-labour ratio, thus allowing land of higher rent to be used more intensively. Second, the treatment of economies of scale is too simple. A production function with increasing returns for a homogeneous consumption good does not shed much light on the structural intricacies of a modern city. Multiple products and trade between towns should be introduced (some first thoughts on how this might affect city size were offered by Lave, 1970). Similarly, indivisibilities in transport should be taken into account.

The first analyses of land use for transport within the CBD were by Livesey (1973) and Sheshinski (1973), reacting more or less simultaneously to a paper by Mills and de Ferranti (1971). The latter examined the problem of land-use allocation for transportation in a simple model in which the urban population and the CBD radius were fixed. The aim of the model was to show how congestion affected the optimal land allocation. Among their results was the striking finding that at the edge of the CBD all the land might be used for transportation. The two other studies showed that this result could be obtained only by assuming that the CBD was fixed, and that this assumption meant a departure from optimality. Livesey shows that in an integrated city with transportation both within and without the CBD "the optimal allocation of land for transportation as we move out from the city centre is a monotonically increasing concave function until the boundary of the CBD is reached when it becomes a monotonically decreasing convex function until the boundary of the city is reached when it is zero" (Livesey, 1973, page 158). If the CBD boundary is allowed to vary, in no case (except the pathological zero-rent case) is all the land used for transportation.

This model is quite restrictive because it ignores the production aspects of the CBD; as in the analysis of Mills and de Ferranti the number of workers (and the city population) is given exogenously. A refinement would be to extend Dixit's analysis by allowing for CBD transport costs, which amounts to much the same as extending Livesey's model (1973) by allowing production conditions to be determined endogenously. A first effort in this direction has been made by Livesey (1976). He ignores what happens in the residential ring in order to concentrate on the CBD. His aim is to develop a relatively realistic model of the CBD, in which profits are maximized in an increasing returns industry, subject to the congestion costs arising from the journey to and from work.

He assumes that one firm produces one commodity under increasing returns to scale. The problem is to maximize the net social benefits of production after taking account of social congestion costs. This means maximizing the difference between the value of output and land (both for production and transportation), labour and capital costs, as well as transport congestion costs (which are a function of the number of commuters). For optimality, conditions of marginal social efficiency have to be satisfied. A solution of this problem involves two crucial choices: choosing the amount of land devoted to transportation at each location, which also determines the land available for production; and deciding the labour force at each location, that is, determining the optimum ratio for land/labour.

Three solutions are considered. The first is the limiting case where all land in the CBD is devoted to production so that there is no intra-CBD travel (this is, in effect, the Dixit model). The solution is easy. The land devoted to transportation is zero, the land–labour ratio is zero within the CBD (workers are only employed at the edge of the CBD) and equal to the area of the CBD border divided by the total labour force at the border, and total transport demand (as measured by the number of commuters) at the border is equal to the city labour force. The result that the factory town employs all the city's labour force at the CBD border but that no one travels within the CBD implies that a plant can be operated as efficiently with all its workers working on the periphery as having them spread throughout the plant. It is hardly satisfactory.

The second solution is to hold factor-input ratios constant at all points in the CBD. This implies that workers are distributed evenly throughout the CBD. The solution is similar to Livesey's earlier model (1973), a feature of which is that some land must be used for transportation everywhere within the CBD. The third case is the most realistic. This involves disaggregation of the production function so that factor proportions are location-specific, although price is still determined by total production and conditions of aggregate demand. Conditions of marginal efficiency have now to be satisfied at each location. At the centre of the CBD higher rents will induce less land-intensive production, which implies more use of labour (and capital) and hence more land for transportation than in the second case. Near the CBD border, on the other hand, fewer workers would be employed and more land would be used in production.

Alao (1976) explains the agglomeration of production in the CBD not in terms of increasing returns but in terms of transportation economies associated with concentration around a centrally-located port or other transport terminal (compare the analysis of Fales and Moses, 1972). This allows him to assume constant returns to scale in production—an analytical convenience that does not make much difference to the analysis. By using the location-specific approach and a simple linear-transportation function (that is, excluding congestion), he shows that land-use intensity within the CBD declines with distance, and that this result is compatible with the operations of a competitive market. However, the introduction of congestion, in the form found in Livesey's model, and the requirement of similar simplifying assumptions, such as a fixed land–labour ratio, result in a rent gradient which is higher than that achieved under competition. As a result, tolls might have to be imposed to convert the competitive into an optimal solution. This analysis is weakened, in Alao's own view, by its neglect of agglomeration economies, defined as economic and cultural benefits. Since knowledge about these is so flimsy, it is not yet possible to explore the question of whether the scale of these benefits is affected by the internal spatial structure of the CBD. Moreover, understanding these benefits may be crucial in explaining the evolution of secondary centres in cities. In Alao's view (1976, page 21), "efforts at constructing those multicentre models are not likely to be very productive until the nature of benefits generated by unipolar urban structures are clearly understood". Despite these pioneering studies, the CBD remains very much an 'empty box' in NUE models of urban spatial structure.

Empirical evidence of increasing returns

Assessment of the hypothesis that cities exhibit increasing returns has been handicapped by data shortages. However, a recent study by Segal (1974) has shed some light on this issue. His analysis is not from the same perspective as the NUE models, and is set up rather differently. For instance, he uses a traditional production function with inputs for labour and capital but not for land. This implies results that cannot be used as an empirical counterpart to models such as Dixit's (1973), which use land and nonland inputs. Also, it provides a striking contrast with other empirical studies of urban increasing returns (Harris and Wheeler, 1972; Edel, 1972), which argue that agglomeration economies and diseconomies are captured in land values[16]. However, Segal takes indirect account of the influence of land by applying cost deflators to the value-added and capital-stock variables, where the higher prices of bigger cities are assumed to be a result of the higher land rents.

Segal also concentrates on manufacturing (in aggregate) and, although this is more compatible with Dixit's approach, it implies a narrow approach to agglomeration economies. With a few exceptions, big cities do not offer comparative locational advantages for manufacturing industries, and urban scale economies are more marked in certain types of service activity. This limitation should be kept in mind when interpreting the results of the study. The main achievement of the study is that it is the first attempt to estimate urban capital stocks [for sixty Standard Metropolitan Statistical Areas (SMSAs) with sizeable manufacturing employment], although the fairly crude approximations involved in the estimates throw doubt on the validity of the results. Nevertheless, it is a useful first step in resolving a critically important question.

The production function estimated has the form

$$Q = AK^\alpha E^{\beta_0 + \beta_i q_i^E} , \qquad (5.2)$$

---

[16] Harris and Wheeler argued that land values would increase proportionately with population in the absence of externalities. They increased faster than population in over one hundred and sixty-five metropolitan areas (indicating net positive economies) in the size range 50000-750000, slower than population in the one to three million range, and proportional to population in cities larger than three million.

where
Q is the real value added,
K is the real capital stock,
E is employment, and
q_i^E is one of a vector q^E of labour quality variables.
This may be redefined as

$$\frac{Q}{E} = B\left(\frac{K}{E}\right)^a E^b , \qquad (5.3)$$

where $b = \beta_0 + \beta_i q_i^E - 1$.

In a cross-sectional analysis of this kind (the study year was 1967), a uniform constant term may impart an upward bias to the exponent estimates. Tests of various city characteristics, however, showed that only heavy reliance on mining was significant. The labour-quality variables included education, sex, race, and age. Only the first two were significant.

The sum of the elasticities in equation (5.3) was 1·017 for the sixty cities, 0·992 for the forty-nine smaller cities and 1·069 for the eleven largest (that is, greater than two million inhabitants) cities. These were not significantly different from unity, apart from the results of the big cities, which were slightly significant (0·10). However, the latter result would have been better if New York could have been excluded as a special case (evidence of decreasing returns in New York is that the residual was negative). In estimates of equation (5.3) the labour term was approximately zero, indicating that there were no increasing returns to labour.

A secondary test was to see if migration flows could be explained by the regression residual patterns. The cities with the highest positive residuals (indicating a more efficient use of productive factors) tended to have high net in-migration rates; those with the highest negative residuals (with one exception, Columbus) lost population. Although large cities attracted population at well above-average rates (mainly because of the success of the one-million-plus cities), the eleven largest cities experienced rather modest in-migration in the 1960s. Several factors contributed to this poor performance: special cases such as New York (a negative net-benefits case?) and Pittsburgh (pollution and poor image); below average education levels; and, with a few exceptions, large female labour forces.

Segal's research leaves the problem of urban economies of scale an open question. His concentration on manufacturing and imperfect data suggest that not too many inferences should be drawn from his sceptical results. Nevertheless, increasing returns remain an important element in NUE models. If population size is to be determined endogenously, it will depend upon increasing returns (which determine income and the numbers employed) and transportation costs (which determine the margin of urban settlement, at least in the CBD workplace city).

An earlier study by Shefer (1973) did not have the benefit of estimates of urban capital stocks. Nevertheless, it was able to obtain estimates of returns to scale at the industry level (that is, localization economies) in SMSAs. Two variants of the CES (constant-elasticity-of-substitution) production function were used. The first, derived from Dhrymes (1965), is

$$w = AQ^b E^c \,, \tag{5.4}$$

where
w is the real wage,
Q is output,
E is labour, and
A is a constant.
The returns-to-scale parameter is $(1+c)/(1-b)$. The second is a variant of Arrow's learning-by-doing model (1962), and is given by

$$\frac{Q}{E} = B w^a Q^{z(1-a)} \,, \tag{5.5}$$

where
B is a constant,
Q is a proxy for cumulated experience, adopted in the absence of capital data, and
a is the elasticity of substitution for labour.
In this case the returns-to-scale parameter is $(1+z)$. These functions were fitted for twenty industries for two periods (1958 and 1963) over a sample of up to sixty-five SMSAs. The returns to scale were quite significant: for all industries in 1963 they averaged 14·38% with the Dhrymes model, and 27·33% with the Arrow model. The large difference between the results cannot be explained. Also, increasing returns were more evident in 1963 than in 1958, and production conditions could hardly have changed much in five years. Finally, the scale estimates for individual industries frequently varied substantially between the two methods and between the two years. The difference between the two equations is mainly that wages are endogenous in the first but exogenous in the second. Which of the two is more appropriate depends on the elasticity of labour supply facing each industry. If labour is highly mobile, wages can be assumed to be exogenous. If each industry faces a limited labour supply, the wage rate will be determined by the amount of labour demanded, that is, endogenously.

There are too many unresolved questions arising from this research to treat it as convincing evidence of agglomeration economies, but it does give some preliminary support for the need to allow for the possibility of increasing returns to scale in urban production functions. As in Segal's study, however, the critical land input is excluded, and in this case ignored.

Cities as public goods: CBD substitutes

To the extent that NUE models have discussed agglomeration economies (more often they have ignored them), the emphasis has been on increasing returns to scale in CBD production. Moreover, the analysis is usually explicitly in terms of manufacturing (the optimum factory town), despite the fact that in most cities all but the smallest types of enterprise have long since decentralized. A more plausible case could be made by disaggregating production and treating the typical CBD business as banking and finance, which need spatial concentration to minimize face-to-face communication costs, and location at the centre (originally, though the argument is less strong today) to draw upon the maximum labour pool (the point of maximum accessibility).

Another approach, frequently adopted in the literature relating to optimal city size, is to stress that the primary advantages of cities lie not in private but in public economies, or in a subset of these referred to as 'urbanization economies'. Much of the analysis has been cost-biased, that is, it has concentrated on the minimum costs of providing public services and has tended to ignore the more nebulous benefits of public facilities. Similarly there has been undue emphasis on the influence of 'public bads' such as pollution, congestion, and crime.

In a recent paper Artle (1973) has extended the theory of public goods to shed light on this question. He defines three subsets of goods: individual goods, collective goods with the property of 'sharing', and a different category of collective goods with the property of 'sharing-and-interaction'. Individual goods are defined in terms of the exclusion by one user of all others, either via the act of consumption or the exercise of property rights. Collective goods involve sharing by two or more users, and the good enters the preference functions of two or more users simultaneously. In many cases, collective goods can be enjoyed only by the complementary application of individual assets (that is, an individual cost). Many urban collective goods have the attribute that the individual can choose whether to consume them or not. The 'sharing-and-interaction' category reflects that for some collective goods the willingness to use them (reflected in the private costs of using them) will depend on the number of other individuals using them. Typical cases are congestion on transport arteries, in parks, or on the beach. This suggests upper bounds on the willingness to use public facilities—but for many facilities there are also lower bounds. In some cases these arise from the costs of producing and maintaining the service; if these are high relative to the number of users the public authority responsible may refuse to supply the service or may withdraw it if it exists. However, consumers may be unwilling to use a facility if there are too few other users. Examples include the empty concert auditorium or sports stadium, the deserted (and in the United States possibly dangerous) park, and the social club or community group suffering from dwindling membership. The potential demand for a service depends in such cases on the scope for interaction, perhaps measured by the number of probable contacts.

In discussing these characteristics of collective goods or assets, Artle uses the cities of the ancient Sumerian, Greek, and Roman civilizations as illustrative examples. Indeed, the facility used in his model is the Greek agora—an open space and a collection of buildings that formed a civic centre as well as a marketplace, the "living heart" of the Greek city and "the daily scene of social life, business and politics". The modern city retains many of the collective assets of the ancient city and more—public places and parks, streets, bridges and subways, cemeteries, concert halls and sport stadiums, transport terminals, and waste-disposal systems. In addition it is a storehouse of knowledge, reflected in schools, colleges, museums, the telephone system, radio, and television. On the other hand, it is weaker than the ancient city in two respects: most of the political and economic power rests elsewhere—in higher levels of government and in the private sector; the modern city is an open system, and the capacity of its services may be strained by the in-migration of households and firms, and by nonresident commuters.

The model is illustrated diagrammatically in figure 5.1. Time is used instead of income, and the individual maximizes his utility by allocating his time. The location of each individual is fixed, and the travel costs incurred in making use of the facility (the agora) are assumed to be the travel time involved. Individuals who live some distance away will need to obtain more benefits from using the facility to justify the extra travel time. Benefits are assumed to be represented by the number of contact possibilities at the agora (assumed constant over time, as is the 'exposure-to-contact' time), which is assumed to be a function of city size. If all

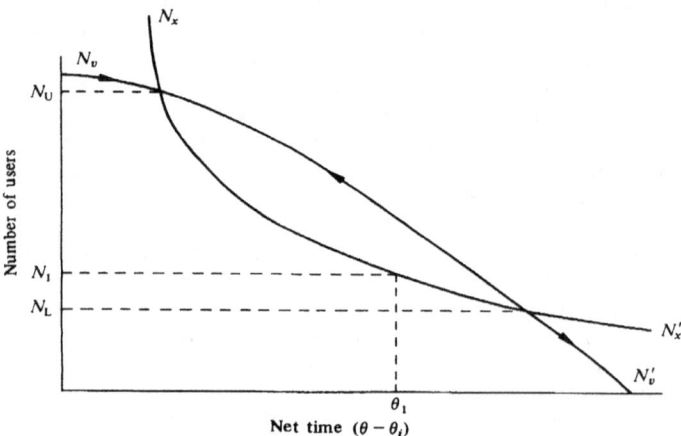

Figure 5.1. Upper and lower bounds for the city as a public good.

individuals have the same preferences, an indifference curve, $N_x N'_x$, can be drawn, which expresses the functional relationship between the number of contacts and the time, θ_i, needed to visit the agora to take advantage of them. An individual living close to the city centre has a large amount of net time, $\theta - \theta_i$, available, and it takes only a small number of contacts to induce him to visit the facility. The further he lives from the city centre, the less net time he has available and the more contacts he needs to use the facility. Since contact possibilities are a function of city size, this helps to explain why residents of big cities are willing to commute longer distances to enjoy public services. The shape of the $N_x N'_x$ function is open to discussion (for instance, Artle suggests that there may be a central kink in it), but the only important points for the analysis here are that there is a limit on how far an individual will travel regardless of city size (the function becomes vertical), and there is a minimum number of contacts needed to make a visit worthwhile even if the travel time is negligible (the function becomes more or less horizontal as net time increases).

Retaining the assumption that residential locations are fixed, a density distribution can be derived which shows the number of individuals for each allocation of net time; this will obviously reflect the residential density distribution of the city. By integration, this density function can be converted into a cumulative frequency distribution, such as $N_v N'_v$ in figure 5.1. This shows the relationship between the amount of available time and the total number of individuals with *at least* that amount of time available. The curve declines, since movement to the right implies fewer and fewer individuals with that amount of net time available. The curve intersects the horizontal axis because each individual has only a finite amount of time available.

The relationships between the two functions illustrate some dynamic aspects of the viability of centrally-located public services (or, more broadly, of the viability of the city centre as a focus of social and cultural activities). Suppose there are N_1 individuals making use of the facility at a given point in time. The corresponding point on the time axis is θ_1; individuals with that amount of net time are indifferent to using the facility and not using it. However, the cumulative frequency distribution shows that there are more than N_1 individuals with at least θ_1 units of net time available. Accordingly more individuals will use the facility. Moreover, since individual utility is an increasing function of the number of users, the increase in users will further encourage others to visit the facility. Thus a cumulative process will develop until the two curves intersect, where the number of users is N_U. This measures the upper limit to the size of the facility. If the number of users exceeded the upper level N_U, some users would be better off—given their preference function, and their location—not using the facility. The decay process would result in a return to N_U, which is a stable equilibrium.

There is also a lower bound in figure 5.1, the intersection of the two curves where the number of users is N_L. This is also an equilibrium, but it is unstable. Any movement upwards or downwards induces further divergences from equilibrium. There is thus a threshold size to the facility. If the number of users falls below the threshold, some of the remaining users will prefer not to visit the facility, and this will induce still others to abstain.

This is a very simplistic model. For instance, the viability of the city centre depends on the interdependence between different types of activity, each with different threshold and ceiling limits. In this case, the bounds of the dominant core activity will determine the city's stability, but it is much more difficult to identify the dominant facility in a modern city than to recognize the critical role of the agora in the city of ancient Greece. Furthermore the utility-maximization framework of this model is even more primitive than that used in NUE models. Nevertheless this model offers an alternative framework for analyzing urban spatial structure, not dissimilar to the recent model of a dispersed city by Beckmann (1976). It solves the problem of how to account for the continued relevance of the city centre in a period of extensive decentralization. It highlights the importance of public services and facilities in urban agglomeration economies, and thereby avoids the monopolistic implications of assuming increasing returns to scale in CBD production (see pages 68-72). It relieves the rent gradient of some of the burden of explaining the structure of the city because it assumes, in effect, a stable spatial structure and transfers the burden of adjustment to the supply of public services. However, this last point reveals its most serious weakness—its assumption that residential locations are fixed. Even if the urban spatial structure changes only sluggishly, the assumption of complete immobility is too extreme and may lead to highly implausible results. Nevertheless, with refinement, the model could play an important role in models of urban spatial structure. In particular, introducing centrally-located public assets may help to defend the assumption of a monocentric city from attacks based on evidence of industrial decentralization, whereas the discussion of accessibility outside the context of the journey-to-work enriches, even as it complicates, the analysis.

A similar approach is adopted in a model of equilibrium in a dispersed city by Beckmann (1976). Instead of a single CBD workplace, he assumes that jobs are distributed in the same way as population—the CBD is assumed away. The substitute agglomerating factor is interaction among households for social and recreational purposes. It is assumed that the transportation cost, or more precisely the time cost of interaction, for any pair of households is proportional to their distance. The scope for interaction is maximized when the average distance between *all other households* and the locating household is minimized. For equal utility to obtain everywhere, those households located at sites away from the maximum interaction zone will be compensated by lower rents.

The monocentric city

The utility function takes the form

$$u = a \log s - \bar{d} + c ,\tag{5.6}$$

where
- s is the space occupied,
- $\bar{d}$ is the average distance to all other households,
- c is consumption, and
- a is a parameter.

Beckmann shows that maximizing this utility function under the condition that it is equal everywhere for all households (given the level of income) generates a density gradient rather different from that of the standard model. In fact, the density gradient is a form of logistic function rather than the negative exponential familiar in urban economics. Densities are much higher at the city centre than at a distance, but the peak is smooth rather than sharp (see figure 5.2). If a given population was being distributed over an area by the two models, the density gradient of the social-interaction model would be much flatter than that of the CBD workplace model.

It is interesting that assuming away the CBD does not prevent the reappearance of a city centre, though in an attenuated form. Centrality has advantages apart from being a locus for jobs. However, it is arguable that the underlying assumption of the model is not very convincing. Households do not want to interact with *all* other households but only with a limited set of households of the same class, status, and interests. This provides a reason why these households might want to live close to each other, but the neighbourhood of agglomeration need not be at the city centre. If the city's households are stratified into classes (by income, status, and race) the basic idea of the model could be used to generate a fairly dispersed density distribution, characterized by individual neighbourhoods that are highly segregated residentially, and each of which has a hillock-shaped density peak (an urban village centre). This is a promising area for future research.

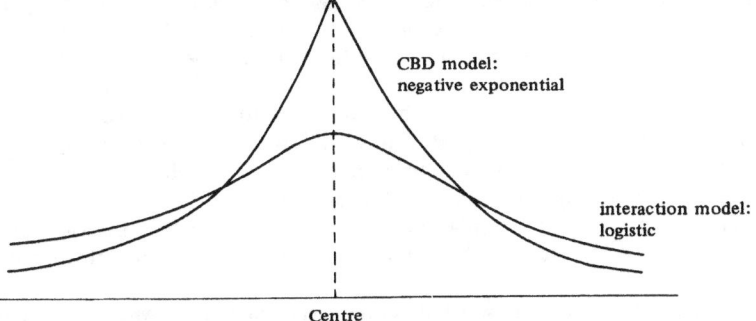

Figure 5.2. Density gradients in the CBD workplace and the maximum-interaction models.

If Beckmann wishes to retain the assumption of maximizing total interaction, it applies best not to households but to certain professional groups (such as lawyers or stockbrokers), where the maintenance of contacts with all other members of the same profession is important for business efficiency. Minimizing communication economies is considered to be one of the more important agglomeration economies for explaining the spatial concentration of certain types of 'centre-oriented' businesses. These are usually found in the CBD. Paradoxically, therefore, although he starts out by assuming away the CBD, Beckmann's model provides—if applied on the lines suggested—a very strong reason for its existence.

Exclusive zoning
Another prediction of the standard urban model is exclusive zoning with "each industry or land use exclusively occupying one concentric zone" (Fales and Moses, 1972, page 52). This is one of the major implications of von Thünen models which are, in a real sense, the antecedent of modern urban economic theory. The other implications—that land-use patterns are the outcome of a competitive allocation process and that the intensity of land use declines with increasing distance from the CBD—are reasonably well established. The results relating to exclusive zoning are much more debatable.

In an interesting study of Chicago in the 1870s[17], Fales and Moses challenge the hypothesis of exclusive zoning. They point out that von Thünen models assume away economies of scale, urbanization economies (for example, in infrastructure), and linkages between firms. They ignore the channelization of transport routes, implicitly assume zero transport costs on intermediate inputs, and—most important of all—they ignore the basic reason for the existence of cities, agglomeration economies. It is not surprising, therefore, that the theory has limited predictive value. There were significant variations in the intensity of manufacturing land use between radial sectors of the city; considerable intermixing among different industries and between industries and households in each ring so that the concept of the exclusive zone was not even roughly approximated; and clear evidence of the existence of satellite employment centres. Employment in nine out of thirteen industries was spread over five or more rings. Capozza (1976) argues that this presents a distorted picture. The same evidence can be interpreted in a different way by pointing out that for eleven of these industries more than 70% of employment was concentrated in two adjacent rings, and the two exceptions were intermediate goods industries (planing mills and millwork) selling output to a wide range of industries. Also, industry was heavily concentrated in

[17] This period was chosen because it follows the great fire of 1871 which wiped out much of the CBD. The fixed capital legacy of the past, which usually fouls up locational analysis of the urban spatial structure, was thus avoided in this case.

the first two rings (68% of employment) whereas population did not peak until the fifth ring. Moreover, the mixing evidence is partly a result of the high level of aggregation—nine rings each one half-mile wide.

Capozza's empirical test is to rework Mills's (1972a) data on the urban-density function to obtain employment/population gradients:

$$\left(\frac{E}{P}\right)_r = \frac{A_E}{A_P} \exp{-(b_E - b_P)r} , \qquad (5.7)$$

where A_E, A_P, b_E, and b_P are parameters to be estimated. This function is negative exponential if $b_E > b_P$. In general, this expectation was confirmed. Also, the slope of the gradient tended to decline over time, suggesting that employment was suburbanizing faster than population. A notable exception was manufacturing in the immediate postwar period when the gradient increased.

The standard model would predict an E/P ratio of infinity within the CBD, and one of zero within the residential ring. The evidence of a gradient gives support to the mixing hypothesis. Capozza explained this distribution (some concentration but also dispersion) in terms of a model of urban location, stressing centre-attraction forces such as agglomeration economies and minimizing transport and communication costs between firms and centre-repelling forces such as lower rents and lower wages (evidence in support of a wage gradient was found from looking at the spatial distribution of earnings for clerical workers in Los Angeles). Capital costs were believed to be neutral between intraurban locations. Because firms differ in factor-utilization ratios and in their need for contacts with other firms, they will also differ in their locational distribution. Capozza suggests that the standard paradigm should be modified to a three-ring model: a CBD where all land goes to commerce and industry; a mixed ring in which goods and service production bids some, but not all, land away from residences; and a peripheral suburban ring totally devoted to housing because no nonresidential demand for land exists at this range. Although a caricature, this is much more plausible than the standard two-zone model. If we allow for a diversity of firms (multiple goods and services) and for the heterogeneity of households, and there is a more or less continuous spectrum of bid-rent functions for both categories, it is easy to obtain *some* residential and nonresidential bid-rent functions which are virtually the same. In this case mixing will occur, especially if it is accepted that there might be some slight discontinuities in the rent gradient, indivisibilities in location, and a degree of randomness in locational choice.

If we now return to the study by Fales and Moses, they argue— reasonably convincingly—that Weberian location theory has been neglected in explanations of intraurban location, and that is is helpful in understanding location patterns in nineteenth-century Chicago. The value of the Weberian approach to the understanding of modern cities is left an open question for further research, in view of the changes in technology, transport improvements, growing importance of services, and the more critical nature of the variations in labour supply and wages that help to differentiate the cities of today from those of the last century. Weberian theory has been ignored because Weber had no theory of household location (population concentrations entered his model merely as cheap labour sites) and failed to determine land rents (though he suggested that increasing land values was the most important deglomerative factor, and rent implications can be derived from his treatment of transport costs on materials and spatial wage differentials).

Industrial location in nineteenth century Chicago can be explained to a considerable extent in terms of the Weberian concepts of transport orientation, labour orientation, materials orientation, and agglomeration economies (a particular form of market orientation). Many of these factors are subsumed in the statements that intraurban freight transport (horse and wagon, which was "slow, undependable and costly") was much more expensive than interurban (train, river, and lake) and it was much easier to move people (by horse railways) than goods within the city. The results were that many industries (bricks, brewing, blast furnaces, etc) were located on the river; railroad terminals attracted many manufacturing industries; materials orientation was more important than market orientation because of the weight-losing characteristics of many nineteenth-century technologies; some activities (for example, commercial banks) located in the Core Zone because this is where the three horsecar lines intersected, and intraurban communications (by messenger) were much more inefficient than interurban ones (by telegraph); population was much more dispersed than were jobs; some activities (for example, savings banks, boardinghouses, and pawnbroking) were oriented towards employment, whereas retail grocers were oriented towards population. It is unclear from this single study how general the scope is for applying the Weberian approach. However, in comparison with the more elegant von Thünen model, it offers a more comprehensive explanation of the variety and irregularity of nonresidential location patterns, much more appropriate for mixed land use than for exclusive-zone cities.

Employment density gradients

In relaxing the assumption that all workplaces are centralized in the CBD, it becomes necessary to choose a technique to measure the spatial distribution of workplaces. One suggestion has been to adapt the density-gradient concept, used in the analysis of population distribution, to deal with the distribution of jobs. The first study to attempt this was by Mills (1972a), which estimated employment densities from the equation

$$(s_r^{-1})^E = (s_c^{-1})^E \exp(-br) , \qquad (5.8)$$

where
$(s^{-1})^E$ is the employment density,
$(s_c^{-1})^E$ is the central-city employment density,
r is the distance from the city centre, and
b is a parameter to be estimated; that is, the slope of the density gradient.

Mills examined broad employment groups (manufacturing, retailing, wholesaling, and services) in eighteen metropolitan areas over the period 1948-1963. He found similar patterns among the study cities. The steepest gradient was for wholesaling, then services, retailing, and manufacturing. There was evidence of increasing decentralization over time. However, the usefulness of the results depends upon the *assumption* that the negative exponential is the most suitable function for employment densities. Fales and Moses (1972) looked at the influence of distance on employment density in nineteenth-century Chicago (using the natural logarithm of distance), but found that other variables such as access to interurban transportation terminals, intersections of intraurban passenger routes, and minimization of communication costs in core areas were more important. This is consistent with the arguments of intraurban location theory which suggest that site choice is a much more complex phenomenon than accessibility to the CBD. Also, as argued by Vernon (1960), with theoretical support from Goldberg (1970), the importance of accessibility varies with size of plant. The central city provides external economies which are beneficial to small firms. For large plants the diseconomies of the central city (high rents and heavy congestion costs) are more burdensome, and they are big enough to internalize many of the externalities. Consequently, returns will tend to be higher at peripheral locations.

The most recent and most detailed study of employment density gradients is by Kemper and Schmenner (1974). Unfortunately the analysis is restricted to manufacturing, a sector which tends to be decentralized and in which interindustry variations may be substantial—especially since the distribution in size of firms varies so much between industries. The study looked at the ten most important industries in five cities over the period 1967-1971, making use of zipcode data to identify locations of individual firms. The results tended to support *a priori* expectations: increasing decentralization over time; employment was more decentralized than the distribution of plants (supporting the hypothesis that big plants tend to be more peripherally located); and few industries were highly centralized, apart from clothing, and printing and publishing.

The more important aspects of the study, however, were its conclusions on methodology. These were so strong as to suggest that the density gradient should be abandoned as a measure of the spatial distribution of industry; in fact, it would be more interesting to have a model of industrial location behaviour that explained the *residuals* of the density gradient.

The estimates of the density gradient were sensitive to the location of the city centre, the measure of central city density, the level of aggregation of the data, and the estimation technique. Cross-sectional results across cities, and even among industries, are unconvincing; a little more justification exists for studies of one city over time.

The negative exponential does not seem satisfactory for manufacturing—a more acceptable function would predict a very rapid decline from high central densities and then a much more gradual decline, not approaching zero until a great distance from the centre. High levels of industry aggregation result in very low estimates of $(s_o^{-1})^E$ and b. The Mills' technique makes almost everything depend on central densities. An increase in $(s_o^{-1})^E$ leads to an increase in b; if the estimate for $(s_o^{-1})^E$ is poor, the approach is invalid. Nonlinear regressions gave different results from log-linear estimates, primarily because the former technique gives more weight to central densities in deriving the fit—and hence yields superior results. The population density and the rent gradients have been important concepts in the formulation of NUE models. This evidence suggests that the employment density gradient is much less useful, and this may slow down the development of dispersed workplace models.

In an early paper in NUE, Niedercorn (1971) developed a model of urban land use that showed how, under certain assumptions, nonresidential demand for land could generate a negative exponential rent gradient. This provides a contrast with the standard model in which residential land use determines the shape of the rent gradient. Niedercorn's approach is capable of two interpretations. The first is to argue that rent is determined by commercial demand, so that households wishing to occupy space in the city have to pay the prevailing rent (this is the assumption behind Niedercorn's theory of residential location). The alternative is to treat the model as a theory of the rent gradient within the CBD (cf Livesey, 1976; Alao, 1976).

Assume that only one commodity is produced in the city and is sold at a constant price, v_o, at the city centre. Although this commodity could refer to any type of industry or activity, a sensible prototype for a city would be to treat it as an activity providing information services. In this case t_r represents the communication (or transport) costs of a firm providing the service at distance r. Output, Q, is a function of labour, E_r, and land, L_r, inputs. Let w be the wage rate [assumed constant throughout the metropolitan area. In accordance with Moses (1962), it is possible to allow for a negative wage gradient; this would not materially affect the results, apart from flattening the rent gradient to some extent]. Π_r are profits, and p_r is rent. Profits at location r are then

$$\Pi_r = (v_o - t_r)Q - wE_r - p_r L_r \ , \tag{5.9}$$

Solving this equation for rent, p_r, we obtain

$$p_r = \frac{(v_o - t_r)Q - wE_r - \Pi_r}{L_r}. \tag{5.10}$$

Locational equilibrium is most easily obtained by assuming perfect competition so that $\Pi_r = 0$ and, assuming that L_r is fixed at each distance r, the urban land market operates so as to maximize rents. Since profits have been assumed to be zero, taking the derivative of equation (5.10) with respect to E_r and equating it to zero yields

$$\frac{1}{L_r}\left[(v_o - t_r)\frac{dQ}{dE_r} - w\right] = 0. \tag{5.11}$$

Assuming a Cobb–Douglas production function of degree one, that is

$$Q = \alpha E_r^\beta L_r^\gamma, \quad \text{where} \quad \beta + \gamma = 1,$$

then

$$\frac{dQ}{dE_r} = \beta \alpha E_r^{\beta-1} L_r^\gamma, \tag{5.12}$$

$$E_r = \left(\frac{\alpha\beta}{w}\right)^{1/\gamma} L_r (v_o - t_r)^{1/\gamma}, \tag{5.13}$$

and

$$p_r = \gamma \alpha^{1/\gamma} \left(\frac{\beta}{w}\right)^{\beta/\gamma} (v_o - t_r)^{1/\gamma}. \tag{5.14}$$

This shows that the form of the rent gradient depends on the shape of t_r. It is plausible to assume that this is concave with declining marginal costs and with distance. A reasonable form might be

$$t_r = k - k \exp(-br). \tag{5.15}$$

If this can be approximated by

$$t_r = v_o - v_o \exp(-br) \tag{5.16}$$

(the cases where $k \neq v_o$ are more complicated though not substantially different), then

$$p_r = \gamma\left(\frac{\beta}{w}\right)^{\beta/\gamma} (\alpha v_o)^{1/\gamma} \exp-\left(\frac{b}{\gamma}\right)r = p_o \exp-\left(\frac{b}{\gamma}\right)r. \tag{5.17}$$

Thus the rent function is negative exponential.

The shape of the transport expenditure function

One of the major drawbacks to progress in urban economic analysis has been the lack of empirical studies to test the central hypotheses. For many years the standard urban model was refined and developed in a vacuum. More recently, some empirical evidence has become available, and it tends on the whole to contradict the predictions of the standard theory.

The particular evidence which led to this indictment refers to the transport-expenditure function.

Transport expenditures have a major role in the theories of urban structure and rent determination, because of the inverse relationship between transport costs and housing costs (rent). The underlying hypothesis of the basic model is that central locations in cities require low transport expenditures and command high rents, while peripheral sites have low rents but involve high transport expenditures. This hypothesis depends crucially on the assumption that all workplaces are concentrated in the CBD. If many workplaces are decentralized, then centrally-located residents may incur much higher transport costs than suburban residents, and this in turn will have implications for the shape of the rent gradient.

In order to examine this, Angel and Hyman (1972a) looked at car-commuting data in the Greater Manchester Area for 1965. Car-commuting costs were measured as time expenditures rather than as money costs. The assumption of radial symmetry was maintained by graphing transport expenditures in relation to distance from the city centre (see figure 5.3), but the assumption of a single CBD workplace was dropped. This means that travel can take place between any pair of points in the metropolitan area, and transport expenditures can be measured either from residences (origins) or from workplaces (destinations). Since journeys-to-work from any given place of residence vary in a decentralized city, the transport-expenditure function measures *average* expenditures. Travel can take place in any direction so that the linear-ray treatment of the standard economic model must give way to analysis in a two-dimensional plane. Finally, whereas economists usually assume that the transport-expenditure function is given and derive the residential distribution, their approach takes the distribution of residences as given and derives the transport-expenditure function as the average expenditures of residents in a given area.

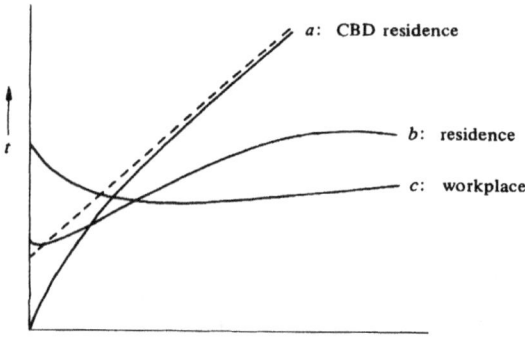

Figure 5.3. Transport-expenditure functions.

Transport-expenditure functions, as obtained from Manchester data, are shown in figure 5.3. Function a measures commuting costs to the CBD from residences at increasing distances. This is the function assumed in the standard economic model, a consequence of the assumption of a single workplace. Transport costs are zero at the CBD, but increase very rapidly close to the CBD because of congestion in the central city, but at a decreasing rate. The function approaches asymptotically to a positively-inclined straight line (the dotted line in figure 5.3). The *actual* (average) expenditures of all residents located at a particular distance are shown by the function b. This is very different from a. It is positive rather than zero at the origin. Initially, it falls within a short range. Since rents and transport costs both have negative gradients within this range, conditions of household equilibrium cannot be satisfied and hence no household will locate close to the CBD (or more precisely, no resident within this range would commute to work by car). The b function then increases generally at a slower rate, though first increasing and then decreasing. Eventually, the function approaches a constant value (presumably because of the general availability of widely dispersed local jobs). Function c expresses the same data in terms of costs from the workplace. Those working in the CBD spend more on travel to work, but the function is U-shaped and employees working at an intermediate distance spend least on commuting (in the Manchester case the minimum point occurs seven miles out).

These comparisons show that the assumption that all workers commute to the CBD grossly overestimates their expenditures on transport. This means that the standard economic model tends to exaggerate the impact of transport costs on rent and on location, compared with a more realistic dispersed workplaces model. The revised transport-expenditure function has implications for rent, density, and wage gradients that conflict with conventional predictions. The standard model, with its indefinitely increasing transport costs, logically implies that rents will eventually become negative. This has to be dealt with either by arbitrarily fixing the city limit or, more satisfactorily, by making assumptions about land-market behaviour that rule out negative rents. However, function b implies that the rent gradient will flatten out when the transport-cost function levels out. This means that the rent gradient will be much flatter than a negative exponential.

Similarly, estimation of the density functions for workplaces and residences showed that the negative exponential did not give the best fit. Workplace densities declined more rapidly than exponentially, and residential densities caved in near the city centre. A gamma distribution of the following type gave a much better fit:

$$s_r^{-1} = ar^b \exp(-cr) , \tag{5.18}$$

where
s_r^{-1} is the density at distance r,
r is distance, and
a, b, and c are parameters to be estimated.
For residences $b > 0$, while for workplaces $b < 0$. In Manchester the estimated functions were

residences: $s_r^{-1} = 1164 r^{0.982} \exp(-0.439 r)$,

and

workplaces: $s_r^{-1} = 4677 r^{-0.451} \exp(-0.298 r)$.

Finally, the evidence of figure 5.3 provides support for the wage-gradient hypothesis associated with Moses (1962) and Muth (1969, pages 42-45). The high cost of commuting to CBD workplaces (function b) suggests that savings accrue to workers if jobs can be found away from the city centre. The number of vacancies and wage levels would tend to be higher at the city centre than at decentralized locations, thereby generating a negative wage gradient. Far away from the city centre, however, commuting expenditures rise from workplaces because employees have to come from greater distances (a consequence of the much lower population densities); the wage gradient may, as a result, turn upwards.

6
The multicentric city

Introduction
The dominance of the assumption of a monocentric city in NUE models is easy to understand. It permits analytical solutions by making the mathematics tractable. The concentration of jobs at one location, and the associated assumption that the workplace is centrally located, allows the two dimensions of space to be compressed into one. This disregard of nonwork trips (either by assuming that shopping trips are part of the journey-to-work or by assuming that shops are distributed spatially in the same way as population) enables the allocation of expenditures between transport and other items in the household budget to be determined relatively easily, as well as allowing a simple treatment of the demand for land for transport purposes compared with competing land uses. Once the assumption of a single workplace is dropped, an early extension is to allow specialization of function among a hierarchy of centres, and this means relaxing the assumption of a composite consumption good, which is a considerable analytical convenience. Once multiple goods are introduced, with specialization among centres, the model has to accommodate intrametropolitan freight shipments as well as deal with a much more complex commuting pattern. It may no longer be possible to obtain determinate solutions, and rent and density surfaces may cease to be smooth and differentiable. In these circumstances, it is hardly surprising that NUE modellers have preferred to compromise with reality and work with the abstraction of the monocentric city.

It is not possible at this stage to develop a satisfactory model of multicentric urban structures. Spatial concentrations in cities are usually explained in terms of the benefits of agglomeration economies, but knowledge of how these forces operate remains very limited. Accordingly only a few eclectic observations as to how such a model might eventually be developed will be offered here. No attempt will be made to suggest solutions to the problem of mathematical intractability.

There are three major questions that need to be answered. Why do secondary centres other than the CBD develop? Where will they be located? What will be their effects on metropolitan structure? Apart from a few casual ad hoc incidental comments, only the last of these questions has received any serious attention. Papageorgiou and Casetti (1971) and Papageorgiou (1974) have carried out some preliminary work on residential spatial structure in a multicentric system. In the latter study, for example, spatial equilibrium is explored in a multicentred city for a continuous income distribution.

The traditional monocentric models have not given much consideration to what induces agglomeration. With a few exceptions (Dixit, 1973; Livesey, 1976; Alao, 1976; see pages 68–72), NUE modellers have treated the CBD as an empty box of no intrinsic interest—merely a dumping ground for commuters. In the few studies that have looked at the CBD as an economic entity, two approaches have been adopted. One is to assume constant returns to scale in production, but to account for the CBD as a predetermined transportation node. The other is to assume an aggregate production function with increasing returns to scale. Usually the CBD is assumed to be aspatial, though Livesey (1976) recently suggested how CBD production might be made location-specific. These analyses are static. If the CBD loses its production monopoly, this can be understood only via a dynamic model. Dixit (1973) argued that the effects of scale economies are eventually offset because of heavy congestion costs for commuters (too much time spent on the journey-to-work). However, in a model permitting the substitution of capital for land in transportation, the problem of heavy commuting could be handled via the construction of a mass-transit system. Thus a more satisfactory solution may be to allow the factor exponents to vary with scale—eventually falling as the scale of output continues to increase. A combination of both solutions may be adopted if the transshipment of goods is assumed to take place. This introduces congestion costs in a large city because high terminal costs rule out highly capital-intensive intraurban freight transportation. Unless the production location is next to a transportation terminal, and the goods are to be shipped out of the city, increasing CBD production will be associated with higher transportation costs because traffic congestion around the CBD affects the efficiency of freight transport by truck.

Lave (1970) outlined a model in which increasing congestion costs eventually erode the agglomeration economies associated with concentration in a single centre, so that it becomes more efficient to divide activities between two or more locations. The first approach discussed here has some affinities with this model. The starting point is to consider a single-good economy in which the advantages of the CBD (assumed to be located in the centre of the urban area) are compared with alternative locations in the city. These advantages, in the simplest model, are of two kinds: agglomeration economies, which arise from spatial concentration of economic activity and/or from economies of scale in production; and transport cost advantages. The first could be achieved anywhere, but occur in the CBD simply because this is the initial site of production. The second type is initially maximized in the CBD because this is the most efficient site from the point of view of transporting goods (or delivering services) to the city as a whole.

In figure 6.1, the agglomeration function A and the transportation function T both vary with the scale of output. Agglomeration economies may be interpreted as a logistic function that initially increases at an increasing rate, but subsequently increases at a decreasing rate towards an asymptotic limit. This shape reflects the plausible hypothesis that agglomeration economies will eventually be exhausted at some scale of output. The transport-cost function of the CBD initially declines because of scale economies in transportation and because of the advantages of the CBD location for serving the city as a whole. Quite soon, however, the function turns upward and the net transportation advantages of the CBD diminish, and eventually become negative when T intersects the horizontal axis. The transportation disadvantages of the CBD then increase at an increasing rate because of congestion costs as CBD output becomes larger.

The net benefits of the CBD, compared to other sites within the city, are measured by subtracting T from A. These benefits first increase then decrease. They are maximized at output Q_{max} where the vertical difference between A and T is greatest. At larger outputs, however, the net benefits fall off rapidly as agglomeration economies grow slowly, whereas congestion costs increase rapidly. At output Q_o transport congestion overwhelms the agglomeration advantages and the net benefits of the CBD disappear.

If this model is interpreted literally, at output Q_o additional production would shift to a new site (beyond Q_o the net benefits of the CBD are negative). This implies a sudden relocation of a kind not observed in real-world cities. The argument becomes easier to accept if two additional points are kept in mind. First, embryonic locations for nonresidential activities will already exist within the metropolitan area (for reasons discussed below), and these will have begun to attract activities. Second, location decisions within the city are not the outcome of an optimization process. Only such a process would lead to a sudden shift. In reality there is a stochastic element in location decisions. For instance it might be reasonable to assume that the probability of a new locator choosing a CBD site is a function of its net benefits[18]. As these net benefits

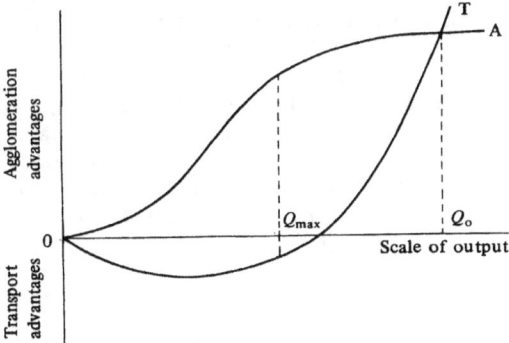

Figure 6.1. Net benefits of the CBD.

[18] If CBD agglomeration economies consist simply of production-scale economies, and not of broader agglomeration and urbanization economies, the market structure should be monopolistic. The monopolist might relocate to a subcentre when output reaches Q_o, as predicted by the optimization model.

become smaller, fewer and fewer new locators will choose the CBD rather than alternative sites. Since locational decisions tend to be interdependent, both rejection of the CBD and agglomeration at another site may become cumulative.

A not dissimilar approach is based on the concept of *agglomeration potential* (Richardson, 1974), which treats agglomeration economies as multidimensional (with congestion costs being counted as *dis*economies) and allows the spatial impact of each to be reduced over intraurban space to different degrees. The agglomeration potential of node j on location i, $^iV^j$, can be represented by the expression[19]

$$^iV^j = \frac{\sum_n w_n^A A_n^j}{\sum_x w_x^d d_{ij}(x)} , \tag{6.1}$$

where
A_n^j is a measure of agglomeration economy (or diseconomy) at node j, and
$d_{ij}(x)$ is the distance between i and j according to variable x (for example,
  transport costs, travel time, and other spatial frictions).
Sets of agglomeration weights and distance weights are represented by w_n^A and w_x^d respectively.

The possibilities of subcentring are explained in this framework by the inclusion of agglomeration diseconomies (negative values) and by the different rates of distance decay for each agglomeration variable. In particular, many diseconomies (for example, certain types of congestion) decay very rapidly with distance near the border of the CBD, but may have disproportionate quantitative weight within the CBD itself. As a result the values of the agglomeration potential for the CBD may be much lower than at certain other locations within the metropolitan area. These higher-value sites are likely candidates for multicentric nuclei. This argument is even stronger in a dynamic context. The relative size of agglomeration economies and (especially) diseconomies within the city change over time, as do the relative values of the variables representing spatial friction. The agglomeration pull of the CBD may thus be eroded over time. The development of several subcentres (rather than merely one, additional to the CBD) may be explained by the critical importance of travel time in the measurement of effective distance[20].

---

[19] There is also a multiplicative version of the concept. The choice between an additive or multiplicative model depends on how certain theoretical and empirical problems are resolved (see Richardson, 1974). The *total* agglomeration potential of the CBD can easily be obtained by integration over the urban area.
[20] Effective distance is the concept developed by Deutsch and Isard (1961) to deal with the multidimensional character of distance.

Intraurban hierarchies
Multicentric spatial structures are easier to explain if the assumption of a single good is dropped, and if the city's economic structure includes different types of activity. The model of a 'factory town' as developed by Dixit (1973) is anachronistic. It might be applied to a nineteenth-century urban area of modest size in a developed country in Western Europe or North America, but it does not bear the slightest resemblance to the structure of a modern city which is usually dominated by service sectors. Manufacturing may be important, but the major scale-economy industries tend to be located on the periphery, whereas CBD manufacturing firms tend to be very small (for example, workshops) and are dependent upon external rather than internal economies.

If several types of nonresidential establishments are assumed, these will differ in their locational requirements: dependence on CBD agglomeration economies, land needs, accessibility to labour pools, proximity to interurban freight terminals, and so on. These differences imply different optimal locations. The advantages of locating inside the CBD will vary among banks, offices, department stores, other shops, public buildings, health and educational establishments, large manufacturing plants, other industrial establishments, and other nonresidential locators. While it would be silly to attempt to take account of this variety in a theoretical model, nevertheless it suggests the implausibility of assuming that all types of enterprises are located in the CBD. The only justification for this assumption is that it simplifies the analysis and permits the use of favoured mathematical techniques. Once this assumption is relaxed, however, multicentric spatial structures are directly inferred. Locational decisions are so interdependent that it takes only one or two major locators to choose proximate sites, more or less simultaneously, to provide the nucleus for a secondary centre. Although this argument is simple when expressed verbally, it is difficult to convert it into a formal model without being much more specific about the characteristics of the key locators.

In the case of many goods and services, the most obvious way of generating multiple centres within a metropolitan area is via a Christaller-Lösch hierarchy model. If we use Christaller's concepts of *range* (the maximum distance over which goods and services might be supplied) and *threshold* (the minimum scale of market justifying production), it follows that the growth of the city in terms both of population and area will be associated with an increase in the number of goods and services and with an increase in the number of centres. Given that market areas are of varying sizes, the locations of many supply centres will coincide. Subsidiary production, distribution, and employment nuclei additional to the CBD will develop in the metropolitan area.

The problem with this analysis is that, in reality, only a small number of sizeable centres are found, whereas the model predicts a large number. However, this contradiction may be reconciled to some extent. Many of the supply centres are superimposed on each other in the centre of the region. The metropolitan area remains core-dominated unless some overpowering factor, such as strong scale diseconomies, intervenes. On the other hand, some service nodes (for example, retail shops, schools) are so small that their distribution over the urban landscape has no tendency to generate a subcentre. In other words, only very high-order goods and services create any significant agglomeration pull. Even if several potential centres emerge from the model, only a limited number will become viable alternatives to the CBD, and the successes will depend on a few major, but stochastic, location decisions which lead to a cumulative process of growth via a sequence of locationally interdependent site selections. Nevertheless the creation of a group of secondary centres (say, four to six) is much more plausible than merely one. The reason is that if the CBD is centrally located, there is no single alternative site capable of sharing efficiently with the CBD the supply of the city as a whole. Six smaller centres, on the other hand, may be suitably located (approximately hexagonally) for supporting the CBD, and ultimately competing with it, during city growth. Christaller-Lösch models start from the assumption of uniformly distributed population. This assumption leads to many more widely dispersed centres than are obtained with a more realistic population distribution. Similar effects result from an uneven spatial distribution of income.

The Löschian version of the hierarchy model is superior to Christaller's for this purpose because it permits flexibility in the number of centres in each rank of the hierarchy. Moreover, in the region as a whole, Lösch's nets of market areas generate sectors that differ considerably in the number of centres they contain. This pattern of 'full' and 'empty' sectors has implications for the probability of emergence of major subcentres. Although Lösch's model was originally applied to the spatial distribution of manufacturing, it can be extended to cover the case of service industries. This is more justifiable when it is recognized that the possibility of excess profits was admitted by Lösch himself: "with discontinuous settlement, the possible size of the market areas and the number of settlements they contain also grows continuously. This, again, makes surplus profits possible. ... Such moderate surplus profits are actually the rule for it would be pure chance if the demand curve in its jumps should still 'just touch the cost curve'" (Lösch, 1954, page 120). Needless to say, the Löschian model becomes even more plausible as the foundation of a subcentring model if it is reinforced by introducing agglomeration economies among unlike activities. Since such economies of urbanization provide the most clear-cut explanation of intraurban nodes, it is unnecessary to force the intraurban central-place models to bear the strain of attempting to explain subcentring alone.

The location of subcentres need not be explained in theoretical terms. By using a hierarchy type model, it might be feasible to predict the 'ideal' distance between subcentres and between these and the CBD, but it would be most surprising if the actual secondary centres developed at their predicted locations. In fact it is more sensible to treat the locations as being determined exogenously or randomly, though subject to certain constraints. For instance, they will tend to be some distance away from the CBD—partly to be outside the central congestion zone, partly to obtain a degree of shelter from the spatial competition of the CBD. Their precise location, however, will be determined by the specifics of the individual city. In some cases, *locational constants* (Richardson, 1973c) will be the most likely sites, especially in large cities. Locational constants are historically-determined locations that have attracted nuclei of population and/or economic activities in the distant past, for reasons which are not relevant to present-day conditions. For example, as the metropolis spreads it absorbs long-standing small towns and village centres that are prime candidates for later subcentres. Other common examples are a suburban education, office, or hospital complex, sites of multiple-mode transportation terminals, and—a more recent phenomenon—suburban shopping malls. These locations attract additional economic activities, new jobs, and population.

Multicentric models
The effects of multicentric cities on spatial structure are much more difficult to analyze than for the monocentric case. It is much easier to tackle this problem with discrete models of the programming type (Hartwick and Hartwick, 1974; see pages 173-177). In the field of continuous models there have been one or two preliminary attempts. Papageorgiou and Casetti (1971) examined the case of identical households in a hierarchy of centres, each order producing a different commodity. It was shown that, to attain equal utility, households are employed in the nearest production centre and make journeys to higher-level centres for services and goods not supplied at the workplace.

The setting is a homogeneous region in which there is a predetermined number of centres. These form an n-order hierarchy, and function as workplaces (of identical jobs) and as distribution centres for goods and services. A higher-order centre supplies all the goods and services provided by lower-order centres plus an additional set that defines its order ($i = 1, 2, ..., n$). Centres of the same order are identical. The population of this region is homogeneous in income and preferences. Each household maximizes its utility function

$$U = U(c_i, s, r_i) , \tag{6.2}$$

subject to

$$u_{c_i}, u_s > 0 , \qquad (6.3)$$

$$u_{c_i c_i}, u_{ss} < 0 , \qquad (6.4)$$

$$u_{c_i c_j}, u_{c_i s} \geq 0 , \qquad (6.5)$$

$$y = \sum v_i c_i + p(r_i)s + T(r_i) , \qquad (6.6)$$

where
c_i is the composite consumption good of order i,
r_i is the distance between the household and the centre from where it obtains goods and services of order i,
s is the quantity of housing space,
v_i is the price of good i,
$p(r_i)$ is the rent at a given location, and
$T(r_i)$ is the total transport costs at that location.

The household will choose to interact with the nearest centres that meet its demands for jobs, goods, and services, and it will always work in the nearest centre. Every residential location in the region is associated with only one vector (r_i).

Assume a household is constrained to live at location (r_i). It will then choose the optimal mix of goods and land that yields a constrained maximum of U when (r_i) is given. Spatial equilibrium will be consistent with this result if no household has an incentive to relocate. A sufficient condition is that the optimal utility level is constant everywhere in the region.

To consider a specific example, let there be one highest-order, n, centre, and let

$$U = \left(\prod c_i^{a_i}\right) s^b \exp\left(-\sum g_i r_i\right) , \quad \text{where} \quad 0 < a_i, b < 1 , \qquad (6.7)$$

and

$$T = t \sum f_i r_i , \qquad (6.8)$$

where
t is the transport cost rate, and
f_i is the frequency of interaction with a centre of order i.

The highest frequency, f_1, refers to the place of employment. This is the lowest-order centre with which the household interacts (this unlikely result is a consequence of the assumptions of ubiquity of jobs and of their identical nature at all centres; a more realistic model would allow for a hierarchy of jobs). If the functions are as given in equations (6.7) and (6.8), the maximum utility level ($\bar{u}$) is obtained from

$$\bar{u} = H\left[\frac{b}{p(r_i)\left(\sum a_i + b\right)}\right]^b \left(y - t\sum f_i r_i\right)^{\sum a_i + b} \exp\left(-\sum g_i r_i\right) , \qquad (6.9)$$

where
$$H = \prod \left[\frac{a_i}{v_i\left(\sum a_i + b\right)}\right]^{a_i}.$$

Let $\bar{u}$ be equal to the utility level consistent with spatial equilibrium ($\bar{\bar{u}}$). Then the spatial-equilibrium rent surface, $\bar{\bar{p}}(r_i)$, is given by

$$\bar{\bar{p}}(r_i) = \frac{b}{\sum a_i + b}\left[\frac{H}{\bar{\bar{u}}}\left(y - t\sum f_i r_i\right)^{\sum a_i + b} \exp\left(-\sum g_i r_i\right)\right]^{1/b}, \quad (6.10)$$

and the equilibrium quantities of land are given by

$$\bar{\bar{s}}(r_i) = \left[\frac{\bar{\bar{u}}}{H}\left(y - t\sum f_i r_i\right)^{-\sum a_i} \exp\left(\sum g_i r_i\right)\right]^{1/b}. \quad (6.11)$$

In this model, locations with prohibitive transport costs have zero rents. Papageorgiou and Casetti prove that the spatial-equilibrium rent surface, $\bar{\bar{p}}(r_i)$, has the following properties:
(1) $\bar{\bar{p}}(r_i)$ is at an absolute maximum in the highest-order centre n;
(2) the values of $\bar{\bar{p}}(r_i)$ in a given order, j, of centres falls as the distances between j centres and higher-order centres increase;
(3) local maxima occur in centres;
(4) there may be some centres that are not local maxima;
(5) if the number of centres is finite, maxima are found only in centres.

More recently, Papageorgiou (1974) has generalized the model to deal with heterogeneous households, in particular, households of different income levels. He explores two types of situation. In the first, called 'abundance consumption', households consume an above-subsistence amount of every commodity; in the second, 'mixed consumption', some commodities are consumed only at a subsistence level. The difference between the two cases, of course, is that in the second case there are additional constraints on maximization of the utility function—a set of subsistence constraints. Since different commodities are supplied in different centres, this constrains the location of less well-off households to be close to the nearest centres supplying the subsistence-level services (these constraints mean that these households have steeper bid-rent functions around such centres). Higher-income households are not so constrained, and income will increase with distance from the critical centres. The result is a determinate income–distance function, rent gradient, and population density gradient. On the other hand, in the abundance-consumption model, the subsistence constraints do not apply. As a result income and location are independent of each other, unless additional restrictive assumptions are introduced: the solution is indeterminate. Even spatial income segregation is less rigid than in the standard models. It probably develops only because of housing indivisibilities which prevent the poor from buying small amounts of housing in wealthy areas (for example, exclusive zoning ordinances).

Similarly, other factors excluded from the model—such as considerations of environmental quality, quality of schools, fear of crime—inhibit wealthy households from moving into low-income neighbourhoods. Without such additions the general multicentric model does not generate predictable location patterns. The indeterminacy is accompanied by other consequences such as irregular density and rent gradients. Since these features are quite important in real-world cities, it suggests that the standard monocentric model is a caricature rather than an elegant, acceptable, and suggestive simplification.

Economics of decentralization

Lave (1974) developed a simplified model of his earlier analysis (1970) of the economics of decentralization, which has a surprising result: "When the cost of commuting rose enough in conjunction with a fall in freight transportation cost to induce a movement away from a single city centre, one immediately noticed a jump to a vast number of centres, rather than a gradual increase to two, and then three, etc." (Lave, 1974, page 57). Of course, as secondary centres develop, the number observed in the real world is not 'vast'. The point, however, is that cities such as Los Angeles, Chicago, and other multimillion metropolises have many centres rather than two or even three. The overproduction of centres by the model is due largely to its simplicity and to the fact that its conception of cities is as a workplace, shopping centre, and locus of production rather than as a complex system for economic and social interaction.

Decentralization will save commuting costs and rents (assumed to be equal to travel costs per mile, t, multiplied by the radius of the city, r_1; thus at the city boundary, rent is zero). Assume that the single city of radius r_1 could be broken up into n cities of equal size (a departure from the hierarchy model for simplification). Further assume that the size of residential sites remain the same (another simplification) so that an average density, s^{-1}, may be assumed regardless of the degree of decentralization. It follows that the radius of decentralized cities can be determined by

$$r_n = r_1 n^{-0.5} , \qquad (6.12)$$

where r_n is the radius of n cities. Total rents and travel costs, C, are given by

$$C = ts^{-1}\pi r_1^3 n^{-0.5} . \qquad (6.13)$$

As the number of cities, n, increases, costs fall very rapidly though by diminishing amounts.

If this were the only element in the situation, it would pay to atomize the city in order to save both on rents and on commuting. The end result would be a widely dispersed spatial system of one-household cities. However, cities exist because of agglomeration economies, so these must be introduced in the model to place a constraint on decentralization.

There are many ways of doing this, such as allowing for public goods subject to economies of scale. Lave prefers to deal with the problem by permitting increasing returns to urban production.

He assumes total demand is fixed at, say, Q units, regardless of the number of commodities. He further assumes that n goods are produced, so that the commodity mix proliferates with decentralization. Production is assumed to be independent. Lave assumes that the benefit of the single city is that no freight costs are involved; purchasers shop on the worktrip without additional cost. If there is more than one city, increasing returns dictate commodity production specialization. However, if demand is the same everywhere, a fraction of output, $(n-1)/n$, must be transported to the other cities. These freight costs are the losses associated with sacrificing the benefits of agglomeration. If the cost of transporting a unit of each of the n goods per mile is t_g, the total freight costs associated with decentralization, T, are given by

$$T = \frac{t_g \bar{d} Q(n-1)}{n}, \qquad (6.14)$$

where $\bar{d}$ is the mean distance between city centres.

If an optimal degree of decentralization exists, it will be found at the value for n which minimizes total costs $(C+T)$, where the additional rent and savings in commuting cost from decentralization exactly offset the additional freight costs (the agglomeration loss), that is where $dC = dT$. Thus the problem is to find the number of cities that satisfies the optimization criterion.

$$Z_{min} = (ts^{-1}\pi r_1^3 n^{-0.5}) + \left[\frac{t_g \bar{d} Q(n-1)}{n}\right]. \qquad (6.15)$$

Lave assumes that the area remains constant with decentralization so that the decentralized cities are circular and tangential to each other. This has the implausible result that there are empty spaces within the metropolitan area in the gaps between the circles. In the example shown below, the more efficient hexagonal configuration is preferred (figure 6.2).

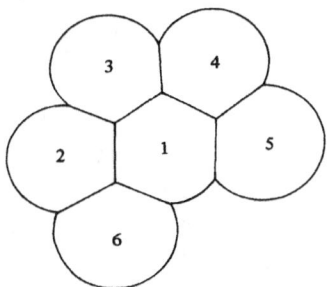

Figure 6.2. A contiguous city with six equal centres.

This means that the area of the decentralized system declines a little as n increases. It also makes the calculation of mean distances a little more complicated. The drawback of this approach is that, given the assumption of a constant average density, the population of the system declines as area declines (see tables 6.1 and 6.2). This lack of realism is not important in view of the unrealistic character of the model as a whole. In the real world, decentralization will tend to be associated with population growth and expanding output, which are not examined in the model. It is not difficult to cope with either a stable or an increasing population by allowing the radius of the decentralized cities to expand or by permitting a change in densities. However, since this would complicate the calculations it is not pursued here. The spatial configuration of a six-centre metropolis, shown in figure 6.2, involves a smaller mean distance than alternatives, that is, pyramid or linear nesting. It is also assumed that cities have common linear boundaries in the two-, three-, four-, and five-centre systems.

Table 6.1 illustrates what happens to relative rent and commuting costs on the one hand and freight costs on the other as a city is broken down into six smaller cities with a radius satisfying equation (6.12). The upshot is that rent and commuting costs decline very rapidly with decentralization,

Table 6.1. Rent, savings in commuting cost, and additional freight costs caused by decentralization.

	Number of centres, n					
	1	2	3	4	5	6
Area ($\times \pi r_1^2$)	1·000	0·972	0·943	0·929	0·921	0·915
Rent and commuting costs ($\times ts^{-1} \pi r_1^3$)	1·000	0·687	0·544	0·465	0·412	0·372
r_n ($\times r_1$)	1·000	0·707	0·577	0·500	0·447	0·407
$\bar{d}$ ($\times r_1$)	0	1·23	1·00	0·98	0·97	0·94
Exports/output $[(n-1)/n]$	0	0·50	0·67	0·75	0·80	0·83
Additional freight costs ($\times r_1 Q t_g$)	0	0·615	0·670	0·735	0·776	0·783

Table 6.2 Total costs of decentralization with variable ratios of freight/commuting cost.

n	Population (thousands)	Cost of decentralization ($ million)						
		$t_g/t = 1·0$	0·5	0·3	0·25	0·2	0·15	0·10
1	314·3	3·14	3·14	3·14	3·14	3·14	3·14	3·14
2	305·4	8·31	5·24	4·01	3·70	3·39	3·08	2·78
3	296·5	8·41	5·06	3·72	3·39	3·05	2·71	2·38
4	292·0	8·81	5·14	3·67	3·30	2·93	2·56	2·20
5	289·4	9·05	4·67	3·62	3·23	2·84	2·84	2·08
6	287·7	9·00	5·09	3·52	3·13	2·74	2·74	1·95

Parameter values: $r_1 = 10$; $Q = 10^6$; $s^{-1} = 10^3$.

though the rate of decline becomes smaller. Conversely, freight costs increase with decentralization, though the increase is relatively modest (with a declining rate) beyond $n = 2$.

It is not possible to identify the optimal degree of decentralization without specifying parameter values for r_1, Q, s^{-1}, t_g, and t. This is undertaken in table 6.2. Since it is the relative freight to commuting costs that matter, t is treated as the numeraire. Table 6.2 shows the results of what happens with decentralization to a city of radius ten miles with an average population density of 1000 per square mile, given varying assumptions about the t_g/t ratio. The results confirm Lave's argument. If the t_g/t ratio is very high, the monocentric city is optimal because it involves no additional freight costs. When the t_g/t ratio declines to a low enough level to justify decentralization (in this case when $t_g/t \leqslant 0\cdot 25$), it is more efficient to have many centres rather than two or three. In the case where $t_g/t = 0\cdot 25$, n nust be at least six before total costs fall below those prevailing in the monocentric city. There are constraints on the proliferation of subcentres in a metropolitan area, but these are primarily outside the scope of this model.

7

More complex residential location patterns

Atypical bid-rent functions
The standard model of urban residential land use started out as a model of a city made up of equal-income households. Subsequently this was generalized to two and then to n residential classes, stratified by income. The prediction in these cases was that the richer classes would live further out than the poor. Although this may be reasonable if the analyst is thinking of the typical North American city, it is rather restrictive. For instance, in many developing countries—such as those in Latin America—the rich tend to live at close-in locations, whereas the poor tend to live on the outskirts of the city in barrios or squatter settlements[21]. Moreover in very large Western cities the residential pattern may show wealthy households living either in the city centre or in distant suburbs, with the poorer households located primarily at intermediate distances. The Manhattan–Westchester County pattern in New York or the Westminster–Ascot pattern in London are not so exceptional that they can be ignored by the theory. What is more, an important tendency to be taken into account in dynamic analysis is a kind of reverse filtering, the invasion of formerly low-income residential areas by the rich (sometimes called 'gentrification').

The residential distribution in developing countries and the core–suburban split of the rich in some (usually big) cities in developed countries can easily be represented in terms of assumptions about bid-rent functions. Figure 7.1(a) shows the standard case where the poor have steeper bid-rent functions than the rich. The common explanation is that it is necessary for the poor to live near work to economize on transport costs, whereas the rich have a high income elasticity of demand for space. If the rich have steeper bid-rent functions, on the other hand, the result is shown in figure 7.1(b), which corresponds approximately to the case of a developing country.

The derivation of two wealthy residential areas is easily achieved by assuming that the bid-rent function of the rich is sufficiently convex to intersect that of the poor twice; this is shown in figure 7.1(c). It implies that the rich either want to live very close to the city centre (to have access to the 'bright lights' or to minimize the time costs of reaching downtown offices) or in the outer suburbs (more space, access to open country, quasirural life-style, etc). The assumption behind the household bid-rent function is that the household is indifferent to location since each point on the function confers equal utility, so that equilibrium is compatible with the split locations.

[21] There may also be some poor living in ghetto-type areas in older parts of the central city. In these cases they usually live in very overcrowded conditions in large houses abandoned by the rich.

An interesting question is what happens when conditions change. Changes in the populations and incomes of both groups (rich and poor) will have repercussions on the territory occupied by each group. The results in the case of the standard model are fairly predictable. However, Hartwick *et al.* (1976) show that whereas an increase in the wealthy population adversely affects the space occupied by the poor (pressing them into a smaller central city area) and reduces their utility, an expansion in the income of the wealthy actually *increases* the utility and space of the poor because the rich suburbanize faster and the poor-rich boundary shifts outwards. A prediction of this kind implies that changes in residential land-use patterns tend to be orderly and harmonious. However, if the rich live both in the central city and in the suburbs (the case of the highly convex bid-rent function *or*, assuming two wealthy classes, one with accessibility-biased and the other with space-biased preferences) an increase in the income of the rich may compress the area occupied by the poor and, *in extremis*, could squeeze them out of the city altogether. Such a

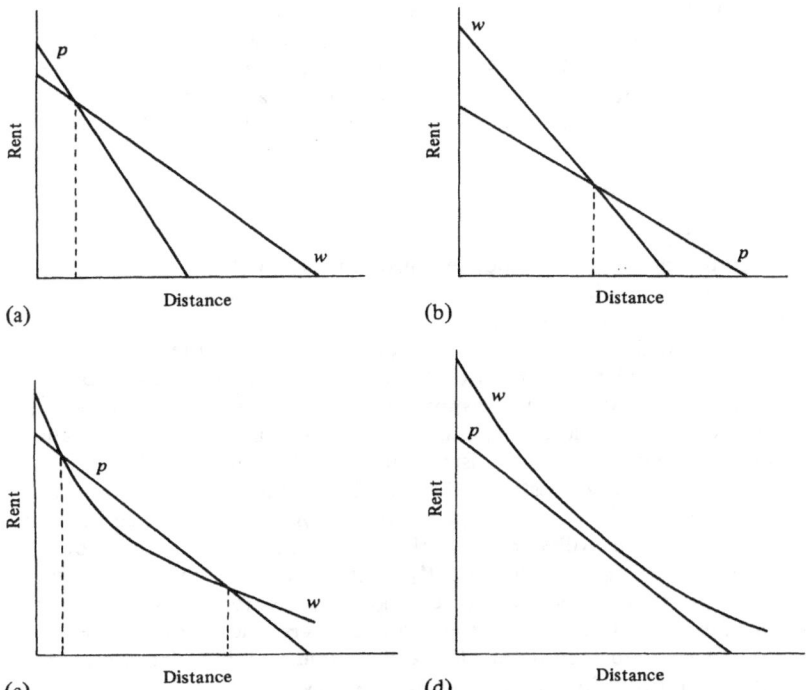

Figure 7.1. Residential distribution patterns. (a) Central-city poor, suburban rich—developed countries; (b) central rich, peripheral poor—developing countries; (c) core rich, inner-city poor, and suburban rich; (d) exclusion of the poor—the one-class city.

case is shown in figure 7.1(d), and results in a one-class city from which the poor are excluded by the rent structure. This type of finding bears a closer resemblance to Marxist analysis than to the neoclassical model: the rich have the market power to exercise their preferences whatever they might be or, more generally, the rich can command space whereas the poor are trapped within it, or even worse, as in figure 7.1(d), excluded from it.

The diagrams of figure 7.1 show deviations from the standard model, but they do not explain them. Some light is thrown on the analysis by thinking of conditions that would rule out multiple intersections of bid-rent functions. If households have the same income but differ in their space preferences (for example, they have the same Cobb–Douglas utility function with the variables housing space and composite consumption, and the housing coefficient is allowed to vary), that is, equal incomes but restrictive variations in preferences, then the functions will not intersect. The same result is obtained if households have unequal incomes but identical preferences, the usual assumptions of the standard model. If households have very different preferences, multiple intersections are possible.

To avoid total indeterminacy it is desirable to analyze this possibility under constrained conditions. A simple case is to consider two residential classes with different incomes. Each is assumed to have a Cobb–Douglas utility function, homogeneous to the degree one, and taking the form

$$U = Ac^a s^{1-a} , \qquad (7.1)$$

where
A is a constant,
c is consumption of the composite consumption good,
s is housing space, and
a is a parameter to be estimated.

It should be noted that distance (or location) does not enter into the utility function, and this implies that time costs—which are functions of income—are not included in transport costs. The simplest assumption is that transport costs from a particular location are the same for rich and poor alike, so that the income distribution may be conceived in terms of income net of transport costs. Heterogeneity of preferences for housing space between income groups can be allowed for by a difference in their $(1-a)$ values. Since utility varies with income, the forms of equation (7.1) for the two income groups are directly analogous to the two-production functions—the two-technologies case of growth theory. There is a very clear and direct connection between double intersections, the bid-rent functions of the two income groups, and double-switching from one technique to another in aggregate-growth theory.

Income and distance

An alternative approach is to introduce distance explicitly into the utility function

$$U = U(c, s, r) \,. \tag{7.2}$$

This is more realistic because it recognizes that different households have different locational preferences, and in addition it allows transport costs to vary with income because of different valuations of travel time. It would be possible to obtain the rich–poor–rich result by stratifying the rich into two groups, one with a negative marginal utility of distance ($u_r < 0$), the other with a positive utility of distance ($u_r > 0$). Some of these problems have been analyzed by Riley (1974). However, the derivation of solutions becomes very difficult.

Muth (1969) has shown that the condition for locational equilibrium with a utility function of the type expressed in equation (7.2) becomes

$$p'_r s_r + t'_r = \frac{u_r}{u_c} \gtreqless 0 \,, \quad \text{as} \quad u_r \gtreqless 0 \,,$$

or

$$p'_r = -\frac{1}{s_r} t'_r + \frac{u_r}{s_r u_c} \,. \tag{7.3}$$

Equation (7.3) implies that p'_r could be positive, even though t'_r is positive, if u_r is positive and the ratio (u_r/u_c) is sufficiently large. This means that in the CBD workplace model, rents might increase with distance. Consequently, Muth argues that "introducing preferences for location ... renders the theory devoid of any empirical content" (Muth, 1969, page 41).

The weakness in this argument is that it assumes that the residential-rent function *must* be negative. The empirical support for this is not as strong or as unanimous as is sometimes argued. Moreover, theoretically, there is a degree of inconsistency between a positive marginal utility of distance and a negative rent function. If households (especially wealthy households) prefer to live a long way away from the city centre and shun central locations, this fact should be reflected in the demand for different sites in the land market. It thus becomes unreasonable to think of rent solely as location rent to be determined by distance from the CBD. A modification suggested elsewhere (see pages 150-157) is to allow for another element in land rent—externality rent—which increases with distance because of favourable neighbourhood and environmental quality characteristics. The movement in rental values over space would then depend on the relative importance of location rent and externality rent. The rent–distance function could be either positive or negative or both (for example, quadratic). In a recent paper, Pines (1975) has shown how Alonso's concept of the bid-rent function can be used to interpret whether or not income increases with distance in different models. The trick is to differentiate Alonso's expression for the bid-rent function with respect to income and then to examine the implications of each model for the variables in the expression.

The slope, dp/dr, of the bid-rent function is given by (see page 15)

$$\frac{dp}{dr} = \left(v\frac{u_r}{u_c} - \frac{dt}{dr}\right)\bigg/s \ . \tag{7.4}$$

Differentiation with respect to income yields

$$\frac{d^2p}{drdy} = \frac{1}{s}\left\{v\left[\frac{d}{dy}\left(\frac{u_r}{u_c}\right) - \frac{\alpha u_r}{y u_c}\right] - \frac{d^2t}{drdy} + \frac{\alpha}{y}\frac{dt}{dr}\right\} \ , \tag{7.5}$$

where α is the income elasticity of demand for housing space. The sign of the right-hand side of equation (7.5) is ambiguous. Whether or not the rich live further out depends on the signs of u_r, $d(u_r/u_c)/dy$, and $d^2t/drdy$. In several models, such as that of Becker (1965) or that of Muth (1969) in the case where all income derives from work, an income elasticity of demand for space greater than unity ($\alpha > 1$) is a sufficient condition to ensure that the rich live on the outskirts of the city.

Thus the location of the rich in the city, and the question of whether they will move further out in response to an increase in income, cannot be determined *a priori* but depends on the relative values of the main parameters. Increases in income will boost the demand for space but they will also raise travel costs because the marginal disutility of commuting depends on income (the value-of-travel-time component). Consider the two-income groups model under conditions of rising incomes. If the income elasticity of demand for space is greater than unity, and if the disutility per mile of commuting is proportional to the wage rate, with the factor of proportionality constant between income groups, the higher-income workers will always live further out. If the income elasticity of demand for space is less than unity, the same result may hold if this demand is not too price-inelastic. However, if the rich value travel time in relation to the wage rate *more* than the poor do, or if the marginal disutility of commuting increases with distance, then the conventional prediction may not hold. In the first case, travel costs per mile will be more expensive for the rich. In the second case, travel costs per mile will be higher for the peripherally-located income group.

Time constraints and environmental quality

Yamada (1972) has developed a model of residential location which takes a much more sophisticated approach to the determinants of residential-site choice than the access-space trade-off model. Although it falls short of a fully determined urban model because the rent gradient and the transport function are assumed rather than derived, it opens the door to a much more realistic type of analysis. Yamada argues that three types of trade-off are involved in the decision regarding residential location: accessibility versus space; space versus leisure; and accessibility versus environmental quality. The main refinements are to introduce time explicitly (see also Evans, 1974) by considering the consumer's allocation of time between work, journey-to-work, and leisure, and to allow for the fact that distance affects utility positively as well as negatively owing to considerations of environmental quality (see also Papageorgiou, 1976b). If accessibility has to be separated from the environmental quality factor, a simple solution is to deal with accessibility in terms of time.

The starting point is the traditional CBD city. A consumer chooses where to live by maximizing utility, u, subject to a time constraint as well as to a budget constraint. The following assumptions are made:

$$t = t(r) , \qquad (7.6)$$

where $t' > 0$ and $t'' = 0$. Money cost, t, is a linear and increasing function of radial distance from the CBD, r.

Travel time, θ^c, is assumed a linear function of distance. Apart from its effect on leisure time, θ^c is also an input with a negative value in the utility function, since the time spent on travel induces dissatisfaction.

$$\theta^c = \theta^c(r) , \qquad (7.7)$$

where $\theta^c < 0$, $\theta^{c\prime} < 0$ and $\theta^{c\prime\prime} = 0$.

Market rent is assumed to be a decreasing function of distance, thus

$$p = p(r) , \qquad (7.8)$$

where $p' < 0$ and $p'' > 0$.

Environmental quality, q^e, is conceived in terms of absence of pollution. It is assumed to be a function of distance because its chief determinant (apart from climate) is density, which is related to distance in a monocentric city on a homogeneous plain (assumption). Environmental influences are an external effect on a household, and create positive utility as a result of increased distance from the CBD. Thus,

$$q^e = q^e(r) , \qquad (7.9)$$

where $q^{e\prime} > 0$.

The household is assumed to have two sources of income: a fixed amount of nonwage income $\bar{y}$, and wage income proportional to working time, θ^w. The latter enters the utility function negatively. Income is defined as

$$y = \bar{y} - w\theta^w , \qquad (7.10)$$

where $\theta^w < 0$ and w is the wage rate per unit of time.

The utility function also contains the amount of housing space, s, composite consumption, c, and leisure time, θ^ϱ. The problem is to

$$\text{maximize } u = u[c, s, \theta^\varrho, \theta^w, \theta^c(r), q^e(r)] , \qquad (7.11)$$

subject to the budget constraint $\bar{y} - w\theta^w - vc - p(r)s - t(r) \geq 0$, (7.12) where v is the given price of the composite consumption good;

the time constraint $\Theta + \theta^w + \theta^c(r) - \theta^\ell \geq 0$, (7.13)

where Θ is the total time available;

a minimum-hours constraint $-\theta^w + \bar{\theta}^w \geq 0$, (7.14)

where $\bar{\theta}^w$ is the fixed minimum standard number of hours;

and nonnegativity constraints $c \geq 0$, $s \geq 0$, $\theta^\ell \geq 0$, $r \geq 0$. (7.15)

The Lagrangian is defined as follows:

$$\mathcal{L}^* = u[c, s, \theta^\ell, \theta^w, \theta^c(r), q^e(r)] + \lambda_1[\bar{y} - w\theta^w - vc - p(r)s - t(r)]$$
$$+ \lambda_2[\Theta + \theta^w + \theta^c(r) - \theta^\ell]$$
$$+ \lambda_3(-\theta^w + \bar{\theta}^w)$$
$$+ \mu_1 c + \mu_2 s + \mu_3 \theta^\ell + \mu_4 r \,. \quad (7.16)$$

The partial derivatives of u are given by u_c, u_s, u_{θ^ℓ}, u_{θ^w}, u_{θ^c}, and u_{q^e}.
The maximization conditions are

$$u_c - \lambda_1^* v + \mu_1^* = 0, \quad (7.17)$$

$$u_s - \lambda_1^* p(r^*) + \mu_2^* = 0, \quad (7.18)$$

$$u_{\theta^\ell} - \lambda_2^* + \mu_3^* = 0, \quad (7.19)$$

$$u_{\theta^w} - \lambda_1^* w + \lambda_2^* - \lambda_3^* = 0, \quad (7.20)$$

$$u_{\theta^c} \theta^{c\prime} + u_{q^e} q^{e\prime} - \lambda_1^*(s^* r' + t') + \lambda_2^* \theta^{c\prime} + \mu_4^* = 0, \quad (7.21)$$

$$\lambda_1^*[\bar{y} - w\theta^{w*} - vc^* - p(r^*)s^* - t(r^*)] = 0, \quad (7.22)$$

$$\lambda_2^*[\Theta + \theta^{w*} + \theta^c(r^*) - \theta^\ell] = 0, \quad (7.23)$$

$$\lambda_3^*(-\theta^{w*} + \bar{\theta}^w) = 0, \quad (7.24)$$

$$\mu_1^* c^* = 0, \quad \mu_2^* s^* = 0, \quad \mu_3^* \theta^{\ell *} = 0, \quad \mu_4^* r^* = 0, \quad (7.25)$$

$$\lambda_i^* \geq 0, \quad (i = 1, 2, 3), \quad (7.26)$$

$$\lambda_j^* \geq 0, \quad (j = 1, 2, 3, 4). \quad (7.27)$$

The equilibrium conditions are

$$\frac{u_s}{u_c} = \frac{p(r^*)}{v}, \quad (7.28)$$

$$u_{\theta^\ell} = \lambda_2^*, \quad (7.29)$$

$$\frac{u_{\theta^\ell}}{\lambda_1^*} = w - \frac{u_{\theta^w}}{\lambda_1^*} + \frac{\lambda_3^*}{\lambda_1^*}, \quad (7.30)$$

$$\frac{u_{q^e} q^{e\prime}}{\lambda_1^*} + (-s^* p') = t' - \frac{u_{\theta^\ell} \theta^{c\prime}}{\lambda_1^*} - \frac{u_{\theta^c} \theta^{c\prime}}{\lambda_1^*}. \quad (7.31)$$

Equation (7.28) is the well-known condition that the marginal rate of substitution between land and the composite commodity must be equal to the ratio of their prices. From equation (7.29) the Lagrange multiplier λ_2^* can be interpreted as the utility produced by an additional unit of leisure time. The left-hand side of equation (7.30) is the ratio of the marginal utility of leisure time to the marginal utility of income, so it measures the money income that has to be sacrificed for an additional unit of leisure. If the consumer works overtime, then, from equation (7.30), $\lambda_3^* = 0$, so that the last term on the right-hand side of the equation drops out. In this case the value of leisure is less than the wage rate.

Equation (7.31) is the locational-equilibrium condition corresponding to the $(-s^*p' = t')$ condition of the standard model. The left-hand side shows the money value of an increase in environmental quality, and a saving in expenditure on location rents from marginal outward movement (the benefits). The right-hand side shows the marginal costs of moving— the money transport costs, the value of travel time, and the money value of the dissatisfaction of additional travel.

The advantage of this model is that it breaks away from the restrictive access–space trade-off that has dominated residential-location theory. Its stress on environmental quality has affinities with the concept of externality rent (see pages 150–157). It also treats the role of time simply but effectively. Its drawback is that it is only a partial equilibrium model of household location. It does not provide a model of urban spatial structure because rent, transport costs, and densities (a determinant of the quality of the environment) are exogenously determined. For instance, the path of the rent gradient would be influenced by travel time and disutility considerations (negatively) and environmental quality (positively). The complexity of the model makes analytical solutions very difficult, and even progress via numerical solutions would require further assumptions.

Towards greater realism

Most NUE models are very restrictive, both in terms of their assumptions and in terms of their reliance on very specific functional relationships. An interesting question, scarcely explored, is how far the models may be generalized. Of NUE modellers, Papageorgiou (1976b) has moved furthest in this direction. Although his model takes the price system as given, apart from land prices, it is general in many respects. He allows for multiple centres, for a continuum of incomes, and for the influence of environmental quality and time on location decisions. Once the model is expressed in very general form, it is then relatively easy to specify more precise versions.

Papageorgiou's spatial environment contains a continuous variable, ϵ, representing residential attractiveness, and a composite index reflecting topography, housing quality, social prestige of areas, and other factors.

It also contains a hierarchy of centres on Christaller–Lösch lines, with the higher-order centres containing all lower-order commodity groups. In policentric space, a small change in locational-equilibrium conditions may imply a long-distance jump. Prices are assumed to depend upon location. This is obviously true of housing, but most NUE models assume that composite commodity prices are constant over space. This is relaxed in this case to obtain a more general treatment of *spatial* consumer behaviour, of which the standard neoclassical price system is a special case (fixed location or zero transport costs).

Within this framework a sequence of models may be examined. These models are summarized in table 7.1. Model I is the nonspatial neoclassical model in which the household optimizes its consumption subject to a budget constraint and given relative prices. Model II, developed by Long (1971), amends model I to take account of space. Prices are now location-dependent: $v = f(r)$ [22]. For any fixed location, $v(r)c - y$ is convex, so that if $u(c)$ is strictly concave there is a unique solution $\bar{c}(r, y)$, which shows that consumption depends upon location as well as upon income.

The limitation of model II is that it assumes that *all* prices are given, whereas most urban models determine at least one price—the price of residential land—endogenously. Land values determine urban land use, whereas the inverse of the land-demand function generates the spatial distribution of population. Separating out land from the commodity sector leads to model III, the simplest form of NUE model. This is similar

Table 7.1. Sequence of household-equilibrium models.

Model	Utility function	Constraints
I	$u(c)$	$vc - y \leqslant 0$
II	$u(c)$	$v(r)c - y \leqslant 0$
III	$u(c, s)$	$v(r)c + ps - y \leqslant 0$
IV	$u(c, s, r)$	$v(r)c + ps - y \leqslant 0$
V	$u[c, s, t(r), \epsilon(r)]$	$v(r)c + \bar{\bar{p}}s - y \leqslant 0$
		$xc - t(r) \leqslant 0$
		$c - \hat{c} \geqslant 0$
		$s - \hat{s} \geqslant 0$
		$\epsilon(r) - \hat{\epsilon}(y) \geqslant 0$
		$u = \bar{\bar{u}}(y)$

Key: u is utility; c is an unknown set of commodities; v is a known set of prices; y is household income; r is location; s is residential land; p is rent; t is time available for consumption; x is a set of time consumption rates; ϵ is residential attractiveness; $\hat{c}, \hat{s}, \hat{\epsilon}(y)$ are the minimum (subsistence) requirements for consumption, land, and residential attractiveness, respectively, the last being income-dependent: $\bar{\bar{u}}, \bar{\bar{p}}$ are equilibrium utility and rent over space, respectively.

[22] See the key to table 7.1 for the definitions of the terms.

to the model used by Muth (1969), the noncongestion model of Solow (1972), and that used for comparative statics analysis by Wheaton (1974b). The main difference is that transportation is included in the general commodity set in the budget constraint, whereas in NUE models it is usual to separate it out, on the ground that travel expenditures are at least as dependent on household location as housing expenditures.

If spatial preferences are explicitly introduced as an argument of the utility function, the result is model IV. This was implicit, at least, in Alonso (1964), but the full implications were missed in early models by focussing on the problem of the distribution of a number of identical households (same income and preference structure) around a single centre. More complex frameworks include a monocentric city of households with different incomes (Beckmann, 1969), identical households in a multicentric environment (Papageorgiou and Casetti, 1971; Papageorgiou, 1974), and a monocentric city of households, with different spatial preferences, that leads to spatial variations in the optimum utility bundle (Mirrlees, 1972; Riley, 1973).

Models III and IV, unlike models I and II, are not closed because there is one price—p, the price of land—that is unknown. To obtain a solution, at least one additional assumption is required. One common assumption is that households should be indifferent to location, as in model III (the locational-equilibrium condition); an alternative is to equate the supply of and demand for land at each location (a common procedure in some versions of model IV). Similarly, in model III location is merely a parameter, whereas it is a decision variable in model IV.

Model V is an outgrowth of model IV. Distance becomes a much more important variable if time is taken into account. If working and sleeping time are treated as institutional and physiological constants, and if the frequency of purchasing a good is independent of quality, then the time available for consumption depends upon location and the speed of travel (the models discussed here do not take account of congestion, which complicates the analysis). Maximization will be subject to a time constraint, with each type of consumption having a corresponding time-consumption rate (x_i, assumed constant). Also, in a policentric environment (assuming no differences in quality or f.o.b. prices), it is rational to interact with the *closest* centres supplying the required commodities. If the physical characteristics of housing affect household location decisions, these may be treated as a component of c, allowing s to refer to land. The introduction of residential attractiveness is important because it constrains locational choice according to socioeconomic status. Minimum requirements for residential attractiveness ($\hat{e}$) increase with income, so that spatial choice narrows as income rises. It is also reasonable to allow for subsistence requirements for goods, $\hat{c}$, and for land, $\hat{s}$.

The generalization to a multicentric environment makes it difficult to apply the 'supply-equals-demand-for-land' condition. It is easier to work with the spatial-indifference condition which implies that, for a given income, the same utility level is obtained at different locations in equilibrium [that is, $u = \bar{\bar{u}}(y)$]. If spatial indifference is assumed, it then becomes reasonable to treat r as a parameter rather than as a decision variable. The latter course would, in any event, involve an intractable system of partial differential equations in the multicentric case. These modifications result in model V, which is much more complex but more realistic than those usually analyzed in NUE. Since the model does not explicitly separate out travel expenditures, the time constraint $[xc - \theta(r) \leq 0]$ becomes the effective spatial-equilibrium condition. Given all other prices v, the unknown price $\bar{p}$ is determined via this condition. Thus, $\bar{p}$ becomes the adjustment mechanism that moves the system to a state of spatial indifference (locational equilibrium). The use of subsistence constraints implies that some households may be too poor to live in the city ($y < \hat{y}$). Moreover, since the locational flexibility of households increases with income—with the above-noted exception about residential attractiveness—subsistence households will be constrained to live at maximum densities in central cities.

Residential attractiveness is a severe complication because proximity to city centres may have a negative impact upon ϵ. This raises the possibility of positive rent gradients (see also Richardson, 1977) or, especially in the policentric environment, highly irregular rent gradients. ϵ may have any continuous spatial distribution, with local effects distorting the smoothness of the rent gradient. This is in addition to the maximum rent peak of the highest-order centre and to the local maxima (minor peaks) found at lower-order centres.

Model V implies a possible break between spatial equilibrium and the spatial distribution of incomes. If we allow for residential attractiveness, a multicentric metropolitan area destroys the predicted pattern of rings of identical households whose income increases with distance. Of course, the real world does include spatial segregation by income, but this is better explained by housing indivisibilities that prevent the poor from entering high-income neighbourhoods, and by residential attractiveness that inhibit the rich from moving into low-income neighbourhoods, than by idealized income rings. Especially in a dynamic framework, changes in ϵ over time and the evolution of policentric urban forms provide a much more convincing explanation of suburbanization. Locational indeterminacy (where the bid rent of any household coincides with the overall rent surface at more than one point) is frequently obtained—for example, with a Cobb–Douglas utility function and where consumption of all commodities is above subsistence level (abundance consumption). The implication that the location decision may be discontinuous even in continuous space is damaging to the susceptibilities of those NUE theorists who are fond of the abstract regularities of the standard model, but it moves the theory of spatial household equilibrium much closer to the real world. Thus Papageorgiou's analysis is of critical significance.

Locational interdependence

Multiple density gradients

A characteristic of NUE models is that they cannot deal adequately with locational interdependencies. These interdependencies are important in understanding the patterns of the urban spatial structure and its dynamics; the attraction of homes near employment centres to low-income workers; residential segregation by income and by race; repulsion between certain types of activity (banks and heavy industry, industry and high-income residential neighbourhoods). The NUE models avoid these problems by treating the location decision as an isolated act determined by preferences, rents, and distance. Since preferences are usually assumed to be the same for everybody, the models generate neat, clear-cut findings such as exclusive zones and smooth income–distance functions. These results distort reality, though they do not necessarily contradict it—otherwise the models would not have been acceptable.

However, observation of the structure of any city reveals that locational interdependencies are important. There is some value in exploring alternative frameworks which can take account of these. One approach is to generalize the analysis of population distribution (density gradients) from the assumption of one population, which is more or less homogeneous, to several (high, middle, and low-income groups; blacks and whites; jobs, workers, and high-income households; and so on). An interesting idea is that these populations will interact with each other according to principles that are measurable and predictable. Many of these interactions might be expressed in the form of 'attractions' and 'repulsions' that have been used by human ecologists to explain certain aspects of urban location patterns. A more formal, but not dissimilar, approach has been developed by Amson (1976). He explores population interactions via two effects; what he calls gravity attractions (coercions) and density-dependent 'location costs'. By making assumptions about the relative magnitudes of these effects, he generates density–distance functions for each population type. Although these are obtained solely by mathematical operations, the models do have economic and social analogues, and it is not difficult to reflect behavioural assumptions in the values of the parameters.

Amson starts by examining the case where only gravity effects occur. By assuming circular symmetry, each population ($i = 1$) can be expressed as a function of density, $s^{-1}(r)$, by integration,

$$N_i(r) = \int_0^{\bar{r}} s_i^{-1}(r) 2\pi r dr , \tag{8.1}$$

where $\bar{r}$ is the boundary at which s_i^{-1} reaches zero (or is determined arbitrarily; for example, the city boundary may be drawn where variations in the density of each population cease to be interesting). The values of the coercions (this is a more neutral term than 'attractions' since it can apply to negative attractions, that is, repulsions) at a distance r depend only on that distance r and the total population $N(r)$ occupying the circular area of radius r. Thus the coefficients are proportional to $N(r)/r$. They may be denoted by k_{ij}, and since there is a coercion between each pair of populations (and in both directions), they form a matrix $\mathbf{K} = [k_{ij}]$. Positive values indicate attraction, negative values repulsion, whereas zero values imply neutrality or indifference. For locational equilibrium within each population, the following i equilibrium equations must be satisfied:

$$\sum_j k_{ij} N_j(r) = 0 , \qquad i = 1, 2, ..., n . \tag{8.2}$$

This simple system is a matrix-vector equation. It has nontrivial solutions if and only if the coercion matrix $\mathbf{K}$ is singular, and this is the case if and only if its row vectors are linearly dependent. However, subject to this qualification, the model can deal with quite complex combinations of attractions and repulsions, with obvious applications to different kinds of urban populations (for illustrative examples, see Amson, 1976, figures 2 and 3).

The weakness of this simple model is that it makes no allowance for economic factors such as locational costs. This may be remedied by introducing the second type of effect. To simplify the mathematics, the two-population case (N_1 and N_2) will be considered. Each population experiences a location cost, p_1 and p_2. Given circular symmetry these costs have a radial gradient, and their value at distance r is given by the appropriate radial derivative, $p'_i(r)$, with respect to r. These costs might be thought of as land rents (though other possibilities, for example, environmental disruption, exist). The greater the value of $p'_i(r)$ the stronger the incentive for the ith population to relocate to less costly sites. The equilibrium equations in this case are

$$\frac{1}{r} \sum_j k_{ij} s_i^{-1}(r) N_j(r) + p'_i(r) = 0 , \qquad i = 1, 2 , \tag{8.3}$$

or, expressing N in terms of s^{-1},

$$\sum_j k_{ij} s_j^{-1}(r) + \frac{1}{2\pi r} \left[\frac{r p'_i(r)}{s_i^{-1}(r)} \right]' = 0 . \tag{8.4}$$

The cost functions p_i may be given exogenously or they may be functionally related to the population densities. If the latter is adopted, the functional relationship might be assumed to have the following form

$$p_i = W_i (s_1^{-1})^{b_{i1}} (s_2^{-1})^{b_{i2}} , \qquad i = 1, 2 , \tag{8.5}$$

which implies that the rate of change in rents is proportional to a linear combination of the rates of change of the densities. The proportionality constants, W_1 and W_2, are the rental coefficients, the constants b_{ij} are the rental exponents, and the rental-exponent matrix is $\mathbf{B} = [b_{ij}]$. The rental exponents determine the relative degree of dependence of the ith cost on the jth density. Substituting for the derivatives $p'_i(r)$ in equations (8.4) and (8.5) yields the following second-order differential equation:

$$\sum_j g_{ij} s_j^{-1}(r) + \frac{1}{r}\left[\frac{r}{s_i^{-1}(r)}\left((s_1^{-1})^{b_{i1}}(s_2^{-1})^{b_{i2}}\right)'\right]' = 0, \quad (8.6)$$

where $g_{ij} = 2\pi k_{ij}/W_i$, so that there is a matrix, $\mathbf{G} = [g_{ij}]$, which may be called the coercion matrix.

Thus the population distribution of each city can be expressed in terms of two coefficient matrices, the coercion matrix $\mathbf{G}$ and the rental-exponent matrix $\mathbf{B}$. Since the analysis can become very complicated in attempts to find general solutions, Amson considers two very simple cases in a two-population system, one where

$$\mathbf{B} = \begin{bmatrix} 2 & 0 \\ 0 & 2 \end{bmatrix} \text{ and the other where } \mathbf{B} = \begin{bmatrix} 1 & 0 \\ 0 & 1 \end{bmatrix}.$$

In both cases the rent of each population depends on the density of its own population but not on that of the other. More complex situations that allow for interdependence are more realistic, but analytically difficult. Also, there is more justification for the unitary rather than the square exponent; if $b_{11}, b_{22} = 1$, this is equivalent to the 'basic' urban equation $p = Kys^{-1}$ (see page 159), where y (income) remains constant. Moreover there is a case for treating the b_{ij} coefficients as parameters. Since there are limits to the increase in rents in response to rising densities, b_{ij} may be expected to decline at very high densities. These problems can be ignored here since the main purpose of the model is illustrative.

When $b_{11}, b_{22} = 2$, $p_1 = W_1(s_1^{-1})^2$ and $p_2 = W_2(s_2^{-1})^2$; and when $b_{11}, b_{22} = 1$, $p_1 = W_1 s_1^{-1}$ and $p_2 = W_2 s_2^{-1}$. In the first example the equilibrium equations (8.6) can be simplified to

$$\left. \begin{aligned} (s_1^{-1})'' + \frac{1}{r}(s_1^{-1})' + g_{11} s_1^{-1} + g_{12} s_2^{-1} &= 0, \\ (s_2^{-1})'' + \frac{1}{r}(s_2^{-1})' + g_{21} s_1^{-1} + g_{22} s_2^{-1} &= 0. \end{aligned} \right\} \quad (8.7)$$

In the second example, the basic equations (8.6) become

$$\sum_j g_{ij} s_j^{-1}(r) + \frac{1}{r}\left[r\frac{(s_i^{-1})'(r)}{s_i^{-1}(r)}\right]' = 0, \quad i = 1, 2. \quad (8.8)$$

These differential equations can be solved by introducing new functions $\rho_i = \ln(s_i^{-1})$; that is, $s_i^{-1} = \exp(\rho_i)$, which transform equations (8.7) into

$$\left. \begin{aligned} \rho_1'' + \frac{1}{r}\rho_1' + g_{11}\exp(\rho_1) + g_{12}\exp(\rho_2) &= 0, \\ \rho_2'' + \frac{1}{r}\rho_2' + g_{21}\exp(\rho_1) + g_{22}\exp(\rho_2) &= 0. \end{aligned} \right\} \quad (8.9)$$

These equations can be solved numerically for a wide variety of G matrices. For analytical solutions, it is convenient to simplify them still further. If it is assumed that the matrix G is singular, that is, that its determinant $g_{11}g_{22} - g_{12}g_{21} = 0$, then equations (8.7) can be simplified to

$$(s_1^{-1})'' + \frac{1}{r}(s_1^{-1})' + s_1^{-1}(g_{11} + g_{22}) + C = 0 , \qquad (8.10)$$

where C is an integration constant.

An even simpler case can be derived by making the additional assumption that $g_{11} + g_{22} = 0$; the two densities then become simple quadratic functions of distance. For instance, if

$$G = \begin{bmatrix} -1 & 1 \\ 1 & -1 \end{bmatrix}$$

and $C = -4$, the density-distance functions for s_1^{-1} and s_2^{-1} are as shown in figure 8.1. This is one of the simplest possible cases: the rental-exponent matrix, B, has been assumed diagonal, the coercion matrix G is singular, and the restriction ($g_{11} + g_{22} = 0$) has been imposed. Nevertheless, the distributions of figure 8.1 are plausible if s_1^{-1} represents, say, manufacturing employment, whereas s_2^{-1} reflects high-level services of the kind usually concentrated in the CBD. To generate population distributions that approximate plausible residential density distributions, rather different assumptions are needed. For instance, if the G matrix is singular for the case where

$$B = \begin{bmatrix} 1 & 0 \\ 0 & 1 \end{bmatrix} ,$$

and if $s_1^{-1}(r)$ is negative exponential, then $s_2^{-1}(r)$ behaves similarly to a gamma distribution. This is consistent with the findings of Angel and Hyman (1972a) in the Greater Manchester Region, where s_1^{-1} refers to workplaces and s_2^{-1} refers to workers' homes.

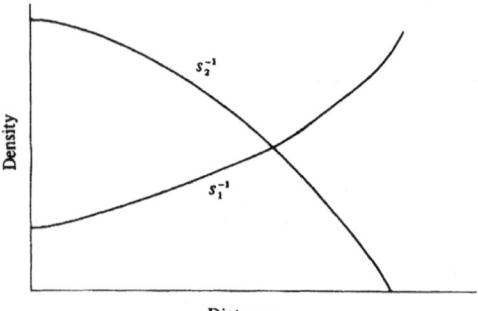

Figure 8.1. Density distributions for two populations.

If the simplifying assumptions of the models outlined above are relaxed, it is clear that this framework can deal with quite complex interactions between different populations and activities (provided that the latter pay rents and are spatially distributed). If there are n populations, with appropriate G values to reflect repulsion between them, and full rather than the simpler diagonal B matrices, quite sophisticated spatial patterns may be generated. Also, this does not imply mere mathematical manipulation. The parameter values assumed for the matrix can be made to reflect deeply researched behavioural assumptions about the locational determinants of different populations and activities. The convincing attraction of this framework is that it focusses attention on the locational interdependence that is believed to be critical in urban location decisions, but which is usually assumed away in traditional theory because it is so difficult to handle. Obvious applications include explanations of the dissimilar distribution patterns of different types of nonresidential establishments, the interactions between workplaces and residences, the agglomeration of like and unlike activities, and the spatial implications of racial segregation behaviour. The surface of this interesting field of inquiry has hardly been scratched.

Racial discrimination and rent gradients

It is possible to amend the standard residential-location model to deal with racial prejudice in the housing market. The starting point is the location of blacks (the argument could apply equally to a low-income group of the dominant race) in an unprejudiced city. There are two main possibilities. If there is no discrimination of any kind (in education, jobs, housing, etc), and if blacks are as productive as whites, households in both races have equal incomes, and if they have the same tastes, the distribution of blacks will be random. This is unconvincing. The more plausible hypothesis is that blacks are discriminated against in education and jobs but not in housing. They will then have lower incomes and different access-space

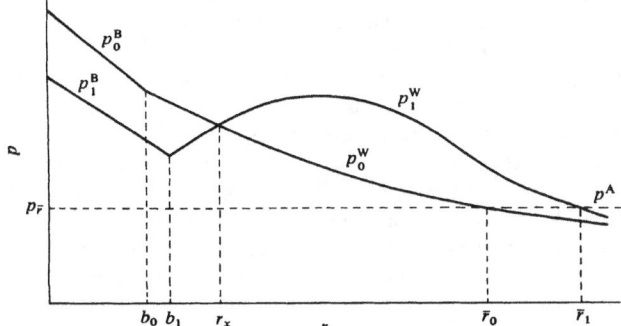

Figure 8.2. The impact of racial discrimination on rent gradients.

trade-offs. On plausible assumptions about the value of the income elasticity of demand for housing, the richer whites will live further out (they have flatter bid-rent functions), and the poorer blacks will occupy the inner city. This type of unprejudiced city has a rent gradient such as that shown by $p_0^B - p_0^W$ in figure 8.2. The rent gradient is steeper in the central city occupied by blacks, and the black-white border is at b_0 where p_0^B intersects the flatter curve p_0^W of the white residents.

What happens if racial prejudice is introduced? White households are both attracted to (more accessibility) and repelled from (closer to blacks) the CBD. This may be analyzed by adopting the utility function that includes distance, but where the relevant distance for the utility of prejudiced whites is the distance from the black-white border, b. Thus the problem is to

maximize $u^W = u^W(c, s, r-b)$, (8.11)

subject to

$$y^W - vc - p_r s - t_r \geq 0 \;, \tag{8.12}$$

where these variables have the same interpretation as before (see page 15). The equilibrium conditions are

$$\frac{u_c}{v} = \frac{u_s}{p} = \frac{u_{r-b}}{p_r s + t_r} \;. \tag{8.13}$$

If we solve for p_r, the result is

$$p_r = \frac{-t_r}{s} + \frac{1}{s}\left(\frac{v u_{r-b}}{u_c}\right) \;. \tag{8.14}$$

Since the second term on the right-hand side of equation (8.14) is positive, rents in the white area fall more slowly than in the unprejudiced city. Indeed they are likely to rise with distance from b. If $p_r > 0$ at b, white rents increase with r until $t_r = v u_{r-b}/u_c$. Beyond this point they fall with increasing distance, and eventually approximate to the rate of change of white rents in the unprejudiced city. The reasoning is that, once white and black locations are far enough apart, the disutility of distance begins to assert itself again. This pattern is shown in figure 8.2 in the $p_1^B - p_1^W$ rent gradient. The p_1 function must cross p_0 at some point (at distance r_x), otherwise all the white households would not be accommodated within the city. The prejudiced whites pay lower rents in the border area but higher rents beyond r_x. The black-white border shifts outwards (from b_0 to b_1), and blacks pay lower rents than in the unprejudiced city because the supply of housing available to them has increased. Assuming that the urban boundary is fixed by the opportunity cost of the use of land in agriculture, p_0^A, the racist city is larger in area than the unprejudiced city, and the urban boundary extends from r_0 to r_1.

Densities are lower in the black ghetto and in the inner white area (between b_1 and r_x) owing to lower rents, but are higher beyond r_x because of higher rents. Densities between b_0 and b_1 will probably be higher because of the filtering from higher-income whites to lower-income blacks. From the point of view of appraisal of NUE models, the interesting point about this analysis (developed by Rose-Ackerman, 1975) is that it provides yet another justification for a positive rent gradient within certain ranges (see pages 150-157).

Some evidence suggests that blacks pay higher rents than whites for equivalent housing. This is explained in terms of disequilibrium. If housing markets are segregated, rapid black in-migration into cities will tend to raise demand for black housing faster than supply, and rents will rise. The model described here is an equilibrium model, hence it is not contradictory to find lower rents for blacks. However, these gains may be small if the price elasticity of maintenance spending on older houses in the ghetto and border zones is high, if costs of demolition and of new construction are high, and if there are large externalities from abandoned or dilapidated dwellings. Even if these conditions do not operate so that blacks gain substantially from lower rents, it is highly dubious that these gains will offset losses from discrimination in the job market. The model cannot be used to promote the idea that the racist city offers blacks economic compensations.

Group-preference models
A characteristic—and arguably a weakness—of most economic models of urban residential spatial structure is their reliance on atomistic behaviour. They emphasize the independent location decision of the individual taken in isolation from those of other city residents. Formally this is dealt with by assuming a utility function without locational interdependence variables. At first sight it may seem surprising that such models result in segregation by income group, but this result is achieved by assuming identical tastes and preferences within groups. An alternative approach, occasionally adopted by sociologists but spurned by economists, is to analyze residential location in terms of groups rather than in terms of individuals. This does have certain precedents—for instance, in Hoyt's radial-sector model and, more recently, in models of racial housing segregation. Nevertheless, this procedure has not been popular, and the frequent aggregation of firms into industries, in microeconomics, has not been carried over into consumer economics by aggregating households into groups.

In the debate about the most relevant theory of residential location, in particular on the relative merits of accessibility ('trade-off') models and environmental-quality and area-preference models, the latter is more consistent with grouping behaviour. Location from the CBD and size of individual site become less important relative to the desire to live near peer groups, in neighbourhoods of higher social status, or—a more indirect objective, though with similar results—to live in areas of high amenity, with good schools and services, clean air, and other attributes of a pleasant social environment. If neighbourhoods are ranked by these criteria, the subset considered in a particular choice of residential site will tend to be spatially dispersed, especially in a large city. The feasible areas may be found at different distances from the CBD because the accessibility factor carries less weight. This tendency will be accentuated in a policentric metropolitan area because most neighbourhoods will be relatively close to a subcentre offering services, and possibly jobs, not dissimilar to those found in the CBD. The result of these factors is that the *individual* site choice will be discontinuous, and much more difficult to predict. Residential patterns will be much easier to explain, however, in terms of group behaviour, with the individual acting as a member of a group rather than as an atomistic decisionmaker.

Grouping in residential urban space may result from three sets of forces. First, there is the possibility of interdependence in housing-demand functions. Households may have a strong preference for living close to other members of their social class, income group, or race. Models based on this type of reasoning are probably the most fruitful of the grouping theories. Second, houses of similar type, size, and cost may be built in a particular area. To the extent that households of the same group have similar housing preferences, this supply may induce intragroup clustering. The two main reasons for homogeneity in housing supply are the lower costs of standardization in building and the role of planning controls, especially lot-size restrictions. Third, grouping may result from market forces in the case where the trade-off between house prices (rents) and commuting costs results in similar locations for individuals from the same income group. In a highly aggregated form, the two-income-class version of the standard model, with its prediction of centrally-located poor and peripherally-located rich, is a simple grouping model. However, it is unsatisfactory if interpreted in this way. This is because it is based on identical tastes. The hypothesis of similar tastes within a group is reasonable if there is a sufficiently large number of groups. Homogeneity of neighbourhoods is found at the very small area level. This requires the analytical framework to be that of a two-dimensional city with a fairly large number of housing groups or classes, stratified by income, race, and social class (or occupational status).

A simple framework for analyzing group residential-location behaviour is via the concept of a *neighbourhood preference premium* (Kirwan and Ball, 1974). The premium is the sum a household is willing to pay for a house of a particular size and quality above the average city price for a house of that type just because that house is located in an area of preferred socioeconomic composition. Although it may be more common for the preferred neighbourhood to be very homogeneous, some households will be willing to pay more to live in a mixed neighbourhood. However,

Locational interdependence 121

if such households are in a minority (a plausible hypothesis), they may be willing to pay more than the ruling price for housing in that neighbourhood, and in this case these households obtain a substantial consumer surplus.

Figure 8.3 shows the preference functions of three different types of household for living in homogeneous neighbourhoods. D is the ratio of households in the locator's class to the total number of households in the neighbourhood; thus, a value of unity implies complete homogeneity, whereas heterogeneity depends upon the socioeconomic composition of the city as a whole (the ratio of households in the reference class in the city to the total number of city households). This is a very simple measure; in a more complex model it may be desirable to use indices of residential segregation. Household H_1 has a weak preference for clustering with members of his own class; the premium, M, he is willing to pay stops increasing when the density of his class in the neighbourhood reaches $0 \cdot 2$. Household H_2 has a much stronger preference for living among members of his own class, but even he dislikes homogeneous neighbourhoods (very intensely) so that the premium he is willing to pay falls off when D exceeds $0 \cdot 5$. Household H_3 is the prototypical residential segregationist with a strong preference for more homogeneity. For values of D below $0 \cdot 75$ he is willing to pay only a negative premium, that is less than the prevailing price. In some cases, his preference curve may be truncated below the horizontal axis, reflecting his refusal to buy a house in an unacceptable neighbourhood at any price.

Of course, a neighbourhood preference model of this kind need not necessarily disturb the optimal location very much from that based on accessibility criteria. This is especially true in the case of very large groups in a monocentric city with a single workplace. In this case the market model and the neighbourhood preference model will tend to produce similar results. On the other hand, if socioeconomic groups are small in size and workplaces are decentralized, the existence of neighbourhood

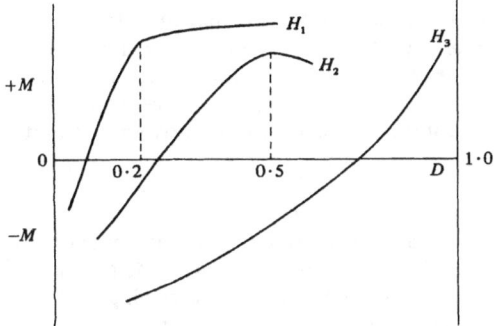

Figure 8.3. Neighbourhood preference premiums.

preferences can lead to significant departures from the optimal locations predicted by individual household models. Moreover, as pointed out by Schelling (1971), since individuals act independently, but satisfaction of neighbourhood preferences depends on the location decisions of others, even a moderate D preference may result in a high degree of segregation. Although equilibria are possible with a low degree of segregation, this will happen only if group preferences are weak and if the groups are large in size, and even in this case the equilibrium tends to be unstable.

Neighbourhood preference premiums may arise from many sources: social status and upward mobility considerations; class identity; high social class of neighbourhood may be a surrogate for environmental quality characteristics and amenities; quality of local schools; minimization of costs of information and search; and reduced risks of capital loss if heterogeneous neighbourhoods are unstable.

The appropriateness of a grouping or area-preference model may be tested in several ways. One possibility is to attempt to measure how much of the spatial variation in house prices or rents can be explained by environmental quality and amenity factors (Richardson *et al.*, 1974a). This is quite difficult owing to the complex nature of income and price effects (Nourse, 1967). Another test is to measure the dispersion of property values within neighbourhoods. After standardizing for house type and size, property values within a neighbourhood ought to be very similar if the group-preference hypothesis is valid, even if the neighbourhood is large in spatial extent so that its houses are located at varying distances from the CBD. This is consistent with Devletoglou's hypothesis of a locational indifference band. Accessibility to the CBD and/or to the workplace may be important only *between* neighbourhoods, not between houses within neighbourhoods. This would imply discontinuities in the rent profile. This is not to argue that accessibility factors are not important within neighbourhoods, but suggests that these refer to access to schools, shops, and open space. A third type of test would be via household surveys, a difficult method that is unnecessary for testing the accessibility models. It is notoriously hard, however, to design surveys in a way that persuades households to reveal their true preferences, especially if their housing preferences include repulsion from a particular ethnic or social group. Since the residential-location decision is the outcome of several, interacting forces it is very easy for a respondent to rationalize a choice in publicly acceptable terms that may, however, hide the true, dominant reason.

Zoning and land values

Very few studies have attempted to evaluate the impact of planning controls on spatial structure and land values. Ohls *et al.* (1974) draw a distinction between two types of zoning—externality zoning and fiscal zoning. Externality zoning, such as the zoning of industrial sites away from residential areas, attempts to achieve a Pareto-efficient land-use pattern via spatial internalization of externalities. Fiscal zoning, on the other hand, typically takes the form of maintaining or achieving a separate and homogeneous tax jurisdiction with the aim of minimizing the tax rate. *A priori*, it is not possible to predict even the sign (positive or negative) of the effects of zoning on aggregate community land values. However, in the case of fiscal zoning, and under plausible assumptions, land values are lower in many US communities as a result of zoning. If zoning throughout the metropolitan area tends to place restrictions on apartment building, the effect on residential land values will be to raise the price in apartment zones and lower it in single-family-dwelling zones. If, in this situation, an individual community zones only single-family homes, it makes its land available only in the lower-priced market, and land values will be lower than without zoning. Yet there are circumstances in which a community concerned with minimizing tax rates ought to favour abandoning the zoning restriction, namely if the consequent increase in land values is large enough to more than offset any increases in public-service costs.

This analysis is subject to the limits of its assumptions. Minimizing tax rates may be merely one of several objectives, among which the desire to maintain a socially homogeneous neighbourhood may be a stronger motive. Moreover, even if restrictive zoning depresses *land* values, much of the land may already be developed, and the zoning restrictions typically enhance the value of existing *properties*, unless the increase in land value from rezoning would be so great as to cover demolition and clearance costs, in addition to paying appropriate market prices for existing properties. Since property owners command more votes than developers, self-interest will normally dictate a vote against rezoning. Even if the increase in land value would satisfy the above condition, locational inertia and a desire to preserve the character of a neighbourhood might still prevail.

The effects of externality zoning are predominantly distributional. If industrial and commercial developments are prohibited in residential areas, there is a welfare gain to the residents but this may be offset by higher transportation and other costs imposed upon firms (assuming that zoning may prevent firms from choosing their optimal locations). Unit land values will increase, but the effect on aggregate land values depends on the consumer elasticity of demand for land, since the increase in land values will reduce consumption. Once again this argument depends on the amount of undeveloped land, since few property owners have scope for reducing their land consumption (owing to indivisibilities) unless their site is large enough to permit subdivision.

Towards dynamics

Compromises

A notable feature of NUE models developed so far is that almost without exception they are static, long-run equilibrium models. In effect they refer to a one-stage allocation problem. They assume that resources are allocated within the city at one point in time—the instant metropolis—or, only a little more plausibly, that adjustment costs are zero. The closest that NUE models in general have come to introducing time is in comparative static exercises, exploring the impact on equilibrium of changes in the numerical values of the main parameters such as income, population, and transport costs, and in showing how rents and densities are affected.

Anas (1976b) has argued that a dynamic model of urban structure would need to take account of two important facts. First, urban infrastructure (for example, roads, housing, nonresidential buildings) is very durable and, once the capital investment has been made, it can be converted only very slowly and at heavy costs. Second, analysis of the housing market would need to recognize that households and suppliers of housing base current decisions on expectations about the future. The second is much more difficult to cope with than the first.

Pines (1976) suggests that the main reason for the retarded development of dynamic NUE models, other than the obvious fact that in a new field static models are usually developed first and progress is subject to lags, is the difficulty of including both time and space in a continuous model, since this requires simultaneous integration over both time and space. The implicit assumption that continuous models are desirable is challengeable, particularly with respect to space: Pines's justification is that "giving up the analytical approach, which requires continuity, the scope for a priori theorems is very limited. Such models [the alternatives] can be used only for simulations based on specific data and sensitivity analysis" (Pines, 1976, page 229). This argument is very much a case for theory for its own sake.

The way out of the dilemma is to compromise with theoretical purity and to allow discontinuities of the kind found in discrete models. Three types of model are possible: (1) models that are discrete both in space and time; (2) models that are continuous in space but discontinuous over time; and (3) models continuous over time but dealing with space discretely. A different approach—briefly discussed below—is to develop nonspatial models of urban growth (typically of the housing market) from which inferences that shed light on changes in the spatial structure can be drawn.

Linear-programming models are the most obvious models of the first type (Mills, 1972b; Ben-Shahar et al., 1969; Ripper and Varaiya, 1974). If development takes place in a sequence of separate stages, it is possible to obtain surprising results. For instance, the city may expand discontinuously, leaving land undeveloped (at least for a time) on the urban fringe and developing exurban sites. 'Leap-frog development' may occur if the disadvantages of high densities are substantial relative to transportation costs, and if the interest rate is neither very high or very low. Low interest rates might permit capital–land substitution to offset the space restrictions of higher densities by erecting high buildings; very high interest rates may prohibit the transportation investments and costlier infrastructure requirements associated with distant, low-density development. Similarly, irregular density gradients may be optimal in a multistage allocation programme; for instance, densities might fall with accessibility—again, because density of development is a function of the factor-price ratios prevailing at the time of building.

Examples of the type (2) model are found in Pines (1976) and Anas (1976b). Pines adapts the well-known Mills and de Ferranti model (1971) to deal with the impact of population growth in a two-stage planning process on the allocation of land to roads (assuming irreversibility of the first-stage allocation in the second phase). In particular, the amended model looks at the net effect of variations in the rate of interest and in the time interval between the two stages of development on the size of the CBD, the size of the inner city (that part developed in the first phase), and the size of the city as a whole. The objective function of the model is to minimize total transportation costs (with residential density determined exogenously). If a Vickrey-type congestion function is used (Vickrey, 1965), then transportation costs in a given ring depend on the number of commuters and on road width (that is, the allocation of land to roads). In the two-stage model the costs of transportation in the inner city are a weighted sum of congestion costs in the two stages.

A simpler version of this model was used to derive analytical solutions, in which congestion costs are ignored and the costs incurred *in the first stage* of the composite input, that are needed to produce transportation capacity to carry traffic in the second stage, represent transport costs in the inner city. Numerical solutions that use the parameters assumed by Mills (1972b) confirmed that the area of the city declined with increases in the rate of interest and lengthening of the time interval between phases. However, the quantitative change was rather small, 1·0% for the city as a whole and 1·4% for the part of the city developed during the first stage.

This model, like the linear-programming models discussed earlier, assumes irreversibility of the initial allocations. This is an extreme case since irreversibility simply means prohibitive adjustment costs. Adjustment costs are usually high but less than infinite and they are difficult to take account of in analytical models. In a model of type (3) (continuous time, discrete space) Hochman and Pines (1973) minimized a sum (over all subareas) of integrals (over time) of housing and transport costs under conditions of increasing marginal-adjustment costs. They were able to show that the dynamics of land-use changes depended on the growth path and the distribution of population, and were different from comparative static results. They were somewhat dissatisfied with the results because they could not obtain a plausible pattern of evolution of land uses under decreasing marginal-adjustment costs, which they regarded as more realistic because of the fixed cost element in demolition. In fact, demolition costs may be very small relative to total building costs, and a more appealing argument for decreasing marginal-adjustment costs is based on psychological grounds associated with locational inertia: the more drastic the change justified by varying economic conditions, the easier it is to make it. Of course, the concept of locational inertia is analogous to that of fixed costs.

Consider again type (2) models (continuous space, discrete time). Anas (1976b) has developed a model which makes some allowance for adjustment from one period to the next in the form of the abandonment of housing stock and the replacement and conversion of housing, though no account is taken of the costs, speed, or path of adjustment. Although Anas does not allow for the expectations aspects of dynamic models and assumes that households and housing suppliers are 'myopic', his model is dynamic in the sense that the housing stock is treated as durable. In fact he assumes that housing is infinitely durable whereas households are perfectly mobile. This implies that if any part of the housing stock is abandoned it is abandoned for economic reasons (rents become negative) not because it wears out physically.

The growth of the city is represented as a sequence of short-run temporary equilibria. Each of these equilibria is constrained by the criterion that utility must be the same everywhere in the city. This is very important for the operation of the model since it explains why inner city rents may decline over time to the extent that they become negative. The city grows as a result of immigration from outside, and the city accommodates the new population by extending the city annularly and building new houses on the urban fringe. In an equilibrium that is short-run, as opposed to long-run, the utility-equalizing rent gradient may decline, increase, or remain invariant with distance depending on parameter values. The basic determinant of a positive short-run rent gradient is rapidly increasing utility over time (owing to appropriate changes in income, the price of capital, and the price of land on the urban fringe). If incomes increase rapidly enough and/or capital costs and land price at the fringe fall fast enough, rents may become negative in the inner city, and the houses there will be abandoned and replaced. If utility increases because of a sharp enough drop in capital and land costs, even though incomes *fall*, the rent gradient will remain negative, and if negative rents appear they will be found at the outermost part of the inner band of the city. A precursor to abandonment may be conversion. This takes place when rents are falling but before they become negative. Also, the city's rental profile will be discontinuous at the border between the existing city and the new ring. This discontinuity is possible in short-run but not in long-run equilibrium.

In long-run equilibrium competitive bidding for sites smooths out the discontinuities. This leads to the conclusion that the short-run city is more suburbanized than the long-run city.

These findings are consistent with the general conclusions of the other attempts at formulating dynamic models. They show that under conditions of growth, land use, densities, and rents may be different from those derived in static long-run equilibrium models. Discontinuous spatial functions, irregular density and rent gradients, and differences in city size—all become possible. Dynamic models are still in their infancy, and as they become more common and more sophisticated they will reveal that the city is a much more complex spatial system than implied by the current NUE models.

Nonspatial models

A breed of dynamic urban models that borrows heavily from aggregate-growth theory is represented by models of the growth of housing stock (Evans, 1975; Muth, 1976). Although these frameworks are nonspatial and hence say very little explicitly about the dynamics of land use, they are not without implications for the spatial structure of cities. For instance, since they show how *average* rents and densities vary, it is not difficult to transfer these into spatial differentials by drawing upon one of the methods for deriving a distance gradient.

The models of Evans and Muth are similar in certain key respects. Both assume a Cobb-Douglas function for housing production subject to constant returns to scale (the large number of small house-building firms suggests that increasing returns are not very important). Both compare a durable housing model with a nondurable model (Evans refers to two types of durable model, one assuming zero foresight, the other perfect foresight)[23]. The durable-housing model is directly analogous to the vintage-capital models of aggregate-growth theory, in which the length of life is an important variable and where different densities provide the housing analogue to different productivities of machines. In order to simplify the analysis, both assume that all variables grow at constant relative rates over time. In other words, they describe a dynamic steady-state equilibrium or 'golden age'.

Since the models are rather similar, it is at first sight surprising that they draw conflicting inferences. Muth argues that it is doubtful whether the refinement of urban models to allow for the durability of housing is worth the effort since urban-residential land-use models that treat housing as nondurable predict land-use patterns, rents, and densities reasonably well. Their main defect is to overestimate average population densities, but this is hardly a case for substituting durable models since these predict

[23] Evans calls the nondurable model Legoland, after the well-known plastic construction toy—Lego. Swan (1956) had described malleable capital as Meccano sets. If Lego is more appropriate for housing, the most suitable fun-name in an urban context is surely Legoville.

even higher population densities than in the nondurable case! Evans, on the other hand, maintains that it would be "dangerously misleading" to rely on comparative-static models. His primary focus is on the behaviour of rents, and he demonstrates that buildings of different ages and heights are compatible with dynamic equilibrium in a growing city, but the rent levels will vary with the rate of growth. Since this result is obtained under golden-age conditions, rents will be even more variable in a city in dynamic disequilibrium.

Superficially, the two models disagree about the behaviour of rents in a growing city, since Muth argues that the durable model predicts lower rents and higher densities than the nondurable case, whereas Evans stresses that rents will be higher when housing is treated as durable. In fact, closer inspection shows that both accept that the result is indeterminate but will depend on parameter values. Moreover Evans concentrates on population growth and ignores income changes, and this makes a direct association between rent levels and densities almost inevitable. Muth, on the other hand, obtains his result via numerical calculations in which the rate of growth of rents is assumed to be slower than the rate of growth in income, whereas the number of households is assumed to grow faster than the land area, but not so much faster that it overturns the rent–income growth ratio. These assumptions about growth are responsible for the counterintuitive result of lower rents associated with higher densities compared with a static or nonvintage model.

Evans generates some interesting results about relative growth rates in the vintage model of housing: the rate of growth of ground rents is faster than the rate of growth of population, which is in turn faster than the rate of growth in rents; the economic life of buildings remains constant as the city grows, but its length is an inverse function of the rate of population growth; if entrepreneurs are myopic, the rate of growth of the city has no impact on the density of dwellings of the latest vintage, and adjustments take place solely by varying the life of buildings or the area being redeveloped. If entrepreneurs have perfect foresight and do not act myopically, some of the results are different. In particular, buildings have a longer life, and the density of new development is an increasing function of the rate of population growth. Of course, both zero foresight and perfect foresight are extreme cases; they provide only a beginning to the analysis of expectations.

This brief survey shows that a meaningful dynamic model of urban spatial structure has yet to be developed. The models developed so far—simplistic dynamicization of clumsy linear-programming models, irreversible investment decisions in two-stage spatial models, analysis of spatial structure from one short-term equilibrium to the next, nonspatial growth models exploring the relationship between urban growth and rents, densities, and other characteristics of the housing stock—are very primitive. There is still much work to be done. Nevertheless, even these preliminary exercises suggest that dynamic models yield qualitatively different results from the standard static models.

The dynamics of density gradients

In view of the immense difficulties in developing dynamic models, there is considerable virtue in dealing with the most simple concepts. An obvious possibility is the density gradient which, when modified to take account of population growth and income growth, sheds some light on the pattern of suburbanization. As is well-known, the density gradient,

$$s_r^{-1} = s_c^{-1} \exp(-br) \,, \tag{9.1}$$

where
s_r^{-1} is the density at distance r,
s_c^{-1} is the central density, and
b is the slope of the density gradient,
may be used to express the urban population distribution if population is a negative exponential function of radial distance in a monocentric city. Mills (1972a), among others, has studied this relationship empirically, and has derived estimates for s_c^{-1} and b for several American cities over time.

If, in a circular city, empty spaces, topographical obstacles, and nonresidential land uses are ignored, then, by assuming a gradient such as that given in equation (9.1) and letting r go to infinity, integration of the densities at each location yields the expression for total city population:

$$N = 2\pi s_c^{-1} b^{-2} \,, \tag{9.2}$$

which can be rearranged to give the following relationship between s_c^{-1} and b

$$s_c^{-1} = Kb^2 \,, \tag{9.3}$$

where $K = N/2\pi$.

To see what happens to s_c^{-1} and b over time, it is necessary to allow the city to grow (or decline). The two main dimensions of urban growth are population growth and income growth. Since urban growth is associated with changes in the spatial structure of the city, growth should affect the spatial parameters, s_c^{-1} and b. For simplicity assume that income grows at a constant rate y', and population grows at a constant rate n. Further assume that an exponential increase in income is associated with an exponential decline in b. This follows from *ceteris paribus* assumptions. Given such assumptions, growth in income will allow households to spend more on travel, and hence to live further out. With a given population N, the density gradient b will become flatter. If income grows at the rate y', b will decline at the rate y' if the income elasticity of demand for space is assumed to be unity. Thus

$$b_\theta = b_0 \exp(-y'\theta) \,. \tag{9.4}$$

Since

$$N_\theta = N_0 \exp(n\theta) \,, \tag{9.5}$$

then
$$s_{c_\theta}^{-1} = \left[\frac{N_0 b_0}{2\pi}\right] b_\theta [\exp(n\theta - y'\theta)] \ . \tag{9.6}$$

Thus central densities and the density gradient are related exponentially (Mogridge, 1974), where the value of the exponent depends on the relative growth rates of population and income. The relationship also depends upon initial conditions as determined by N_0 and b_0. This may be compared with Bussière (1972), who derived a linear relationship empirically with the form

$$s_c^{-1} = f + gb \ . \tag{9.7}$$

He found that the value of f was negative for large European cities, but positive for American and some smaller cities. A linear relationship with $f < 0$ (the Parisian case) is possible only if n becomes smaller relative to y' over time. This is probably correct in European big cities given the demographic dynamics of Western European economies.

It is commonplace in urban economics to use changes in the density parameters, s_c^{-1} and b, as measures of suburbanization. The simple dynamic formulation described here enables a more direct link to be drawn between this measure of suburbanization and two of its chief determinants—population change and income growth. Given the assumption that income growth rates are related to the rate of change in the density gradient, and given estimates of N_0, s_c^{-1}, and b, it would be possible to derive some conclusions about the relative rates of growth of population and income even in the absence of income data.

The simplicity of this model is a warning not to place too much strain on it. It has, at least, three major limitations. Population growth and income growth are not independent, as the framework assumes, but are closely interdependent. This interdependence can only be examined satisfactorily via a simultaneous model (or a simulation model) of urban growth and of the dynamics of urban spatial structure. Second, there is a missing link in any analysis of decentralization that looks at population densities and income but ignores rents (land values or house prices). Even if the NUE models had achieved nothing else, their value would be justified by the attention they focus on the rental gradient as a control variable in analysis of urban space. Third, an empirically observed characteristic of the nature of suburbanization in developed countries is the close interconnexion between decentralization of population (residences) an employment (workplaces). There has been an unsettled controversy about whether or not jobs have followed people or vice versa. This relationship is masked in the simple model discussed here, though there are implicit links between income growth and structural and locational changes in economic activity.

10

An optimum geography

Introduction
Most NUE models, and a high proportion of research in urban economics generally, have focussed on analysis of the urban economy in isolation from the rest of the economic system. They have, in effect, been analyzing 'closed cities', which are characterized by the maximization of utility (utility endogenous) for a predetermined population (population exogenous). A few studies, such as that of Mirrlees (1972), have analyzed the 'open city', where the city population is allowed to change by migration. The typical approach in the open-city model is to assume that utility is exogenous (the level obtainable elsewhere), and that population is endogenous and adjusts so as to equalize utility everywhere. But the interdependence of the city with the wider spatial system is almost always conceived in terms of interaction between the city and its surrounding rural hinterland—a kind of dual-economy model. The spatial distribution of population and the economic activity outside the city are not explicitly considered, and the analyst confines his spatial theory to the treatment of intraurban space. Ideally, however, the focus should be expanded to explain the spatial distribution of the system as a whole—the determination of optimum geography in a true sense. There is a burgeoning literature on the theory of the distribution of city sizes. However, this has had negligible links with intraurban analysis. Also, city-size-distribution analysts have frequently ignored the problems of the spatial distribution of cities. Since space in general and transport costs in particular are constraints on the mobility of goods and population, the size of any urban concentration is interdependent with its location in space relative to other cities.

Another strand of NUE analysis that is of some relevance here is the theory of the multicentric city (see pages 89-101). Although this theory assumes a number of centres and then derives an equilibrium population distribution (so that the number of centres is not endogenous), nevertheless within this limitation there are analogies between this branch of theory and optimum geography. The major difference is that if the context of the latter is the national space economy, the hierarchies where households live and work are partitioned subsets within the national settlement pattern. Additional functions are needed to achieve equilibrium among subsets, and the mechanism for this is interregional migration flows.

Mirrlees (1972) is the only NUE theorist to deal explicitly with the question of optimum geography, in the sense of maximizing a social-welfare *function* for the national population (urban and rural) subject to constraints of production and land area. Beckmann has written extensively on the distribution of city sizes and on the problems of the equilibrium of residential density, but has never attempted to integrate them within the same model. Starrett (1972) has analyzed one aspect of the problem, the increasing returns characteristic of urban production, but has not looked at the problem of the dimensions of residential location. This model assumes identical cities with no interurban trade. Each city can be described in terms of an aggregate production function subject to increasing returns, whereas transportation services are produced in each city under constant returns. Land is used as an input into both production and transportation. In the absence of externalities there is an optimum degree of increasing returns and, if land is homogeneous and the rural sector is self-sufficient, this is equal to the ratio of the total differential rents (urban minus rural) to the value of total output.

Niedercorn (1974) has focussed on the problem of allocating resources (especially land) between urban and rural uses, initially in a 'one-city-in-one-region' context. However, if there are external diseconomies of scale, it soon becomes optimal (in the sense of maximizing total net product) to develop K cities. Given the assumption of a unique production function, these cities will be of equal size (Niedercorn measures size in terms of radius rather than population, a consequence of his concentration on land use). They may be promoted via the national (or regional) government by using a high tax rate to discourage continued growth in the initial city "by throwing the costs of external diseconomies back on the producing sector" (Niedercorn, 1974, page 12). Niedercorn recognizes that introducing a hierarchy of functions leads to a hierarchy of cities, and that attempts to restrict the size of the largest cities (on external diseconomy grounds) may impair the efficiency of those high-order functions subject to the greatest scale economies. However, this may not be a serious planning problem. If the higher-order function serves the total national population, presumably it can do this regardless of the spatial distribution of population. A less primate city size distribution should only imply additional transport costs to service the redistributed population. Those should not have marked deleterious effects unless the activity is operating at the very margin.

The Mirrlees model

In order to obtain results for optimum geography as an extension of his intraurban model, Mirrlees starts out from a set of drastically simple assumptions. These include: identical tastes for all households; a single commodity; labour is the sole input into production, and both capital and land are ignored (even as an input into agriculture!); zero transport costs for commodities. Land and transportation do not enter the model, except for housing and moving people. Production possibilities are represented by a unique production function,

$$Q = H(N), \qquad (10.1)$$

for a single plant employing N people at one site (a city). The rural population live and work in the same place, though each household is

An optimum geography

spatially dispersed, and rural output per head is $h = H'(0)$. The production function H is convex within some range of N; after all, increasing returns are presumably the reason why cities exist.

The problem is to maximize total utility

$$KU(C, N, A_N) + (P - KN)U(c_a, s_a, 0, s_a^{-1}), \qquad (10.2)$$

where
K is the number of towns,
C is the aggregate consumption of a population of a town,
A_N is the area of a town,
P is the national population,
c_a is the rural consumption,
s_a is the rural housing consumption,
s_a^{-1} is the rural density (Mirrlees assumes that low neighbourhood density confers a utility that is different from a large lot size), and
0 represents the fact that distance does not enter into the utility function of rural dwellers (in Mirrlees' urban model, distance is an argument in the utility function).

This welfare function is maximized subject to the production constraint

$$KC + (P - KN)c_a \leq KH(N) + (P - KN)h, \qquad (10.3)$$

and the land-area constraint,

$$KA_N + (P - KN)s_a \leq A, \qquad (10.4)$$

where A is the land area of the country as a whole.

The first-order maximization conditions are

$$u_C = u_{c_a}, \qquad (10.5)$$
$$u_{A_N} = u_{s_a} - s_a^{-2} u_{s_a^{-1}}, \qquad (10.6)$$
$$u_N = u_a - u_C(c_a + H' - h) - u_{A_N} s_a, \qquad (10.7)$$
$$W - Nu_N - Cu_C - A_N u_{A_N} + [H - NH'(N)]u_C = 0; \qquad (10.8)$$

where W is maximum welfare.

Equation (10.8) is obtained by varying K and N together in such a way that KN remains constant. This equation can then be manipulated in combination with the corresponding intraurban optimality condition to yield

$$NH' - H = \int (p - p_a) r \, dr - \int (x^{s^{-1}} - x_a s_a^{-1}) r s^{-1} \, dr, \qquad (10.9)$$

in which the rural rent $p_a = u_{s_a}/u_{c_a}$ (where v, the price of the composite consumption, is the numeraire), and $x^{s^{-1}}$ is an environment density subsidy (or tax) per acre to correct for the externality effects associated with low densities (households have a preference for low neighbourhood densities, but as individuals they are powerless to alter the density of the area in which they live). The expression

$$x_a = s_a^{-1} \frac{u_{s_a^{-1}}}{u_{c_a}}$$

is the 'base rate' per capita of the environmental subsidy that applies to the rural resident (who lives and works in the same place). In the intraurban context the environmental externality correction is needed to convert a competitive residential-density equilibrium into an optimum, and the chosen instrument $x^{s^{-1}}$ is a subsidy (or tax) on commuting.

The first three maximization conditions are, in effect, boundary conditions demarcating the town from the country. Equation (10.5) states that everyone should have the same marginal utility of consumption whether in cities or in rural areas. Equation (10.6) means that the rural land rent should be equal to urban rent on urban boundaries. The third condition, equation (10.7), implies that the utility derived from labour income should be the same in both urban and rural areas. Since urban productivity is higher than rural productivity, this condition may require redistribution in the form of transfers of goods from the town to the country (or possibly *vice versa*).

The fourth condition, equation (10.8), determines the optimum size of the town. The city expands until the difference between marginal and average urban product is absorbed by the excess of urban land rent over rural land rent (minus the excess environmental density subsidy—in a simpler model without externalities this may be ignored). This is similar to Starrett's optimality condition discussed above. The same point has been made by Barr (1972): "Whatever the benefits derived from agglomeration, land rents through the bidding of space will rise to eliminate them. Competition will push city size until there are no net benefits to be gained by any household" (Barr, 1972, page 94). A similar argument has been advanced by Edel (1972): the net benefits of urban agglomeration to the individual are defined as average benefits minus average costs $(\bar{B} - \bar{C})$. However, many of these benefits are monetized in the land market and are absorbed by urban rents. Rent per capita is defined as $(\bar{B} - \bar{C})(1 - \beta)$, where the parameter β measures the proportion of benefits remaining with the individual. However, β can be eroded through competition, and the optimum city size from the point of view of optimality conditions in the economy as a whole is the size at which total rents are maximized, marginal benefits equal marginal costs (MB = MC), and the excess of $\bar{B}$ over $\bar{C}$ is fully absorbed by urban rent (that is, $\beta = 0$). Thus Mirrlees's optimal city size condition is a familiar one in the literature. Urban rents are the equilibrating factor that equalizes the net benefits of urban and rural living. The excess urban rents should, in fact, be redistributed to the rural areas in the form of goods. The amount to be transferred to rural areas is equal to aggregate production, minus the cost of land occupied by the town at rural rents, minus aggregate consumption of goods in the city (transportation expenditures are treated as an element in urban land rent).

Because of nonconvexities in the urban production function, there may be multiple equilibria satisfying the first-order conditions, especially equation (10.8). Thus the optimum city size may not be defined uniquely. There will be a global optimum, but to obtain this would require additional assumptions. Mirrlees assumes that the land-area constraint is not critical so that circular cities may be assumed. This is reasonable for a large country, but might need modification for a small one.

Mirrlees deserves the credit for being the first NUE analyst to attempt to extend the intraurban model to an interurban context. However, as this writer has argued elsewhere (Richardson, 1973a), the analysis suffers from several defects, some of them more serious than others. Probably the weaknesses can be traced to Mirrlees's emphasis on intraurban equilibrium with the optimum geography extension added almost as an afterthought. Had the primary focus been city-size distribution, a quite different set of initial assumptions might have been made, and the analysis would have run on other lines.

Some of the assumptions were made to simplify the model, and could be relaxed without too much difficulty. For example, reliance on a single production function leads to the totally unrealistic system of equal-sized cities, and the assumption of several different production functions does not disturb the maximization conditions. However, were this to be done, a question that might need an answer is why production functions differ among cities. There are many explanations, but most of them involve further departures from the simplicity of the model. For instance, given multiple products the product mix might differ among cities, and indeed an important subset of city size distribution theories (central place and industrial hierarchy models) is predicated on this very hypothesis. Another possibility might be productivity differentials owing to interurban variations in access to technical innovation, but this would require an explicit discussion of interurban space to explain the spatial frictions on innovation diffusion. Mirrlees's model cannot resort to these explanations because it assumes one commodity and it ignores interurban space. The size and efficiency of a city are, in part, a function of its location relative to other cities. It is paradoxical that a theoretical framework explicitly developed to deal with behaviour over space within cities should forget about space when the focus of the analysis is shifted outside cities. The zero transport costs and the one-commodity assumptions are unfortunate, since the result is to rule out two major reasons for the size distribution of cities.

Mirrlees argues that his assumption that neither land nor capital is an input into the production process is merely a simplifying procedure that is easily relaxed. That may be so, but some awkward implications follow

from the assumptions. A truly general equilibrium model would want to
satisfy conditions of marginal efficiency both between and within sectors.
But in Mirrlees's model, no sector uses any capital; industry and agriculture
use labour but no land; housing and commuter transportation use land
but no labour; and commodity transportation uses no inputs of any kind.
This lack of consistency is unsatisfactory, to say the least. The assumption
that agriculture does not use any land is particularly restrictive since in
many NUE models the opportunity cost of urban land in agricultural
production sets the limit on expansion of urban areas. The substitute
explanation for rural rent in Mirrlees's model, that rural residents like space,
is much less convincing. The neglect of capital (though very common in
NUE models) makes it difficult to determine the lot size and the road-width
requirement per person endogenously (the solution with regard to lot size
is to treat housing consumption simply as a perfectly divisible demand for
land alone and to specify the road-width requirement exogenously).
Ideally the amount of land used per person depends upon the choice of
transportation technology, and this in turn depends upon the capital-land
ratio in transportation [of the NUE modellers only Capozza (1973), and
Riley (1974) have explicitly discussed the role of transportation
technology]. Similarly, residential site size is a function of the density of
development, and this depends upon the capital-land ratio in housing.
The introduction of capital allows the space requirements of households
and transportation to be determined by economic considerations, namely
capital intensity, rather than by physical or technical considerations. Of
course, if capital is permitted within cities then it must be allowed in the
economy as a whole. This implies yet another equilibrium condition in a
more comprehensive model—the equalization of returns to capital between
cities, between urban and rural areas, within cities, and between sectors.

Variable production functions
Tisdell (1974) has explored the question of optimal city sizes within the
framework of a very simple model, but one which relaxes the Mirrlees
assumption of identical aggregate production functions. If these functions
are identical and concave it is optimal to distribute the total national
population equally among the available cities. If production functions are
strictly concave but not identical, then unequal size cities will usually be
optimal. Population will be distributed among cities so as to equalize the
marginal product of all settlements where settlement is justifiable (the
model assumes zero transport costs, and is inherently spaceless).

However, the case where the functions are strictly concave is implausible.
As suggested above, a typical urban production function will contain a
convex portion (increasing returns) up to some population level before it
becomes concave. This complicates the analysis even in the case where all
production functions are identical. If we assume no constraints on the
number of sites, cities of optimal size (in the sense of the population size

that maximizes per capita income) result only in the situation where the total population is exactly divisible by the optimal city size. In most cases there will be a residual population, and this may be located in a smaller city, or in one or all of the other cities (thereby expanding their size beyond the optimum), or all the cities may be of equal size but less than the optimum. Depending on circumstances, all these solutions may be compatible with maximization of per capita income in the nation as a whole. If the number of available sites is an effective constraint (that is, if $P > K\bar{N}$, where P is the total population, K is the number of sites, and $\bar{N}$ is the optimal city size), then all cities will be of the same size but larger than the optimum in the case of identical production functions. When production functions are different, on the other hand, maximum overall income may not only require different city sizes but also different levels of production per capita. This is because of a constraint on the number of the most highly productive sites combined with marked concavity of the production functions of these cities at population levels beyond the optimum.

The implication of these arguments is that a migration process that takes place so as to eliminate interurban differences in per capita income (output) may not lead to an optimal national-settlement pattern, except in the special case where urban production functions are strictly concave and identical. With nonidentical, but concave, functions the more productive cities will tend to become overpopulated. With production functions containing convex segments (even if they are identical and there are no constraints in the number of sites), spontaneous migratory processes may result in an equilibrium in which there are large cities above the optimal size and small cities below the threshold size (that is, before the production function displays convexity). This is especially likely if there are only a few big cities, but a large total population to be distributed.

This may be illustrated with a simple diagram. Figure 10.1 shows a production function, $Q = H(N)$, identical for all cities, characterized by initial linearity, then convexity, and ultimately concavity. The classical optimal city size, $\bar{N}$, is found where the marginal output is equal to the average output per head, that is where the linear ray through the origin, OF, is tangential to the production function at B. More generally, the slope of the production function at any city size measures the marginal output per head for that city size, whereas the slope of the linear ray that intersects the production function at any city size measures average output per head. For small city sizes (below the threshold city size at which increasing returns are obtained) the average and marginal per capita outputs are equal. At city sizes smaller than N_1, this identity holds because of the assumed initial linearity of the production function. If the linear segment of the production function OA is projected outwards as the ray OG so that it intersects $H(N)$ at C, it is clear that average per capita output is higher at city sizes below N_1 than at city sizes beyond N_2.

However, city sizes within the range N_1 to N_2 are preferred to cities smaller than N_1 because they offer higher average per capita output. If the total population is large and the number of big cities sufficiently small, settlements within the range N_1 to N_2 may grow beyond the optimum, $\overline{N}$, but not beyond N_2. Rather than join a city larger than N_2 a migrant will receive a higher income in very small cities (below N_1). An equilibrium is possible in which big cities have reached N_2 but there is a random distribution of small settlements below the threshold size N_1. There may be occasional breakthroughs of small settlements into the N_1-N_2 range, but randomly rather than optimally. Thus market forces may result in a suboptimal equilibrium in which some cities are too big while others are too small. A possible weakness in this simple but instructive analysis is its essentially static nature. It assumes that population transfers between cities merely move these cities along their production functions. However, since nonmarginal changes in population size affect market-demand opportunities, input ratios, and other determinants of production possibilities, the likelihood is that there will be consequential *shifts* in production functions. This may disturb the conclusions, but probably in the direction of making a distribution of optimal city sizes even less likely.

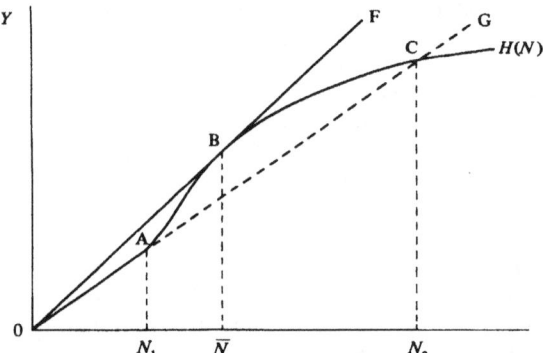

Figure 10.1. Production function approach to optimal city size.

Alternative approaches

A model of the distribution of city sizes that, with extensions, could be made consistent with NUE intraurban models was put forward by Rashevsky (1951). Although Rashevsky himself was a mathematical biologist, his model is pure neoclassics. Migrants move between cities in pursuit of higher income and, given assumptions about marginal productivity, equilibrium is achieved when productivity per head, y, is equalized between cities (and incidentally between urban and rural areas), thus

$$y_1 = y_2 = y_i = \overline{y} . \tag{10.10}$$

An optimum geography

Rashevsky suggests that the generalized income function has the form

$$y_i = f(P_i, N_i), \qquad (10.11)$$

where P_i is the total number of people living in cities of size N_i.
The total number of cities is given by

$$K = \sum \frac{P_i}{N_i}. \qquad (10.12)$$

To solve the problem of the distribution of city sizes the income function of equation (10.11) requires specification. Rashevsky adopted a simple form according to which urban productivity was assumed to be a function of city size and of the characteristics of population distribution of the settlement system (total national population, size-class population, and the total number of cities). This yielded an equilibrium distribution of city sizes satisfying equation (10.10).

Although the model is quite easily reconciled with equilibrium in NUE models, since the latter ensure intraurban locational equilibrium whereas the Rashevsky framework generates an interurban locational equilibrium, a problem is whether the resulting city-size distribution is consistent with the empirical facts—namely, a lognormal or Pareto distribution. As pointed out above, both Mirrlees (1972) and Starrett (1972) assumed identical urban areas by using the same production function. This implies a distribution of equal-sized cities, and of course this conflicts with reality. So how can a lognormal city-size distribution be generated? One solution is to argue that urban production functions are themselves lognormally distributed. This might be plausible by drawing an analogy between business firms and cities, so that the lognormal distribution so characteristic of industrial establishments may also apply to urban areas. Alternatively, and perhaps more satisfactorily, the urban production function may be assumed to reflect several influences that combine multiplicatively (for example, access to natural resources, supply of entrepreneurs, transportation advantages, a high rate of savings, age of settlement, amenities, topography). A corollary of the Central Limit Theorem is that if a random variable, y, is the multiple of independent random variables $(x_i = 1)$, the distribution of y approaches a lognormal distribution as the number of component variables, n, increases, provided that there is no single dominant variable $\left(y = \prod_{i=1}^{n} x_i, \text{ is lognormal} \right)$.

A quite different method of generating a lognormal distribution is via a multisector model in which industries are distributed among cities hierarchically (Tinbergen, 1968). There are more sectors with minor scale economies than those that enjoy large-scale economies, so that there are many more cities that specialize in lower-order sectors (although in applying Tinbergen's model to manufacturing, one-of-a-kind functions tend to be concentrated in the service industries, for example, national government, corporate finance headquarters). Although large-scale-economy sectors may, empirically, tend to pay somewhat higher wages than lower-order sectors, theoretically maximized income per head may be the same in all cities. In any event, as argued below, given the existence of mobility costs, unequal incomes may be consistent with equilibrium. However, the adoption of a multisector model has repercussions on the intraurban system. For instance, it is then unsatisfactory to leave the CBD as an 'empty box'. The location of nonresidential activities becomes a relevant objective of inquiry, whereas the need for interurban trade means that freight, as well as people, are transported within cities.

At first sight the requirement of equal income per head for interurban equilibrium is somewhat restrictive. For instance, interurban (interregional) moving costs may be far from negligible. The appropriate adjustment is to convert the equilibrium condition to equalization of incomes net of mobility costs ($y_j - t_{ij} = y_i$, where $y_j > y_i$). The easiest way of doing this is to assume that migration is financed via a perpetual loan of which the interest cost is deducted from income. A more serious concern is that, empirically, incomes differ between cities, and between cities and rural areas, and by margins wider than those implied by migration costs. In particular, urban incomes are much higher than rural incomes and, within the city-size distribution, real incomes are strongly and positively related to city size (for a review of the evidence see Richardson, 1973b, pages 47-69). Since these income differences have persisted over long periods of time, with no marked tendencies to converge, it is difficult to argue that the interurban system is moving towards equilibrium. One possible explanation is that the system is subject to intermittent shocks and impulses that keep it in a permanent state of disequilibrium. A much more interesting explanation, however, is that *substantial* income differences are consistent with equilibrium.

How can this be true? The clue is to remember that equilibrium is obtained within a city not by maximizing income but by maximizing utility. In the open form of an NUE model, equilibrium utility is determined exogenously and is equal everywhere (this need not hold in a closed-form solution; see pages 65, 66, 131). Similarly, the equilibrium requirement among cities refers to equal utility not equal incomes. In the intraurban model in which public goods and externalities are introduced as arguments of the utility function, maximized equal utilities will imply unequal incomes. In the interurban context the same analysis may apply, and Rashevsky himself argued that his model could be expressed in terms of equalizing 'satisfaction'. Hoch (1972) and Tolley (1974) have suggested a 'compensating-payments' hypothesis based on the argument that incomes have to be much higher in big cities to compensate workers who live there for negative externalities (pollution, congestion, noise, stress, and exposure to crime). These income differences are consistent with equilibrium because, after correcting for externalities, they imply equal utilities. So a more appropriate equilibrium condition in the optimum-geography model is equal utilities between cities, within cities, and between urban and rural areas. Although rural incomes are much lower than urban incomes, rural residents are compensated by low housing costs and zero (or very small) commuting costs. Thus the condition of uniform utility is quite compatible with very irregular income contours both at the macroscopic (national space economy) and microscopic (intraurban) levels.

Evans (1972) has used the economic theory of clubs to develop a theory of the distribution of city sizes. If the coalitions (cities) that result form a stable hierarchy, it can be argued that locational-equilibrium conditions are satisfied. However, Evans's analysis applies to firms rather than to households, and a hierarchy develops because of the assumptions that market opportunities and input costs vary with city size. Similar results were obtained by von Böventer (1970) with the use of a profit-maximizing model. Different firms have varying needs for agglomeration economies. Assuming that the costs of public services rise with city size, a firm whose profits are increased by reaping agglomeration economies will be willing, and can afford, to pay more for public services, and still maximize profits. These models have several problems. Additional variables or assumptions are needed to show how these processes that relate to the choice of city size lead to a city-size hierarchy that approximates the lognormal, Pareto, or rank-size distributions. Moreover, as von Böventer recognized, if households choose where to live (the interurban choice) on the basis of income opportunities, living costs, consumption externalities, and preferences of life-style, whereas business firms make their decisions on the basis of business agglomeration economies, tax rates, and costs of public utilities, there is no automatic compatibility between the two sets of decisions. Too many households may want to live in cities where there are too few firms, or too many firms may want to locate in cities where there are insufficient households. Presumably the solution here is an equilibrating adjustment process in the labour market in which wages fall in the 'excess-household' cities, thus slowing down or reversing the in-migration process, whereas labour shortages in the 'excess-firm' cities raise labour costs and lead to revised location decisions in favour of cities where labour supply is more plentiful and/or wages are lower. However, to assume that everything works out in the end requires an act of faith in the perfections of the labour market and in the lack of inertia in relocation and migration decisions. Finally, expressing the model in terms of business firms makes its adaptation for the purposes of optimum geography difficult, partly because Evans assumed away the problem of intraurban business location by regarding (on typical NUE lines) all employment to be located at the city centre, whereas von Böventer made merely passing reference to industrial decentralization, but mainly because a general equilibrium model of intraurban nonresidential location has yet to be developed. So long as the intraurban location of employment is determined exogenously, this avenue is not very promising as a solution to the problem of optimum geography.

A more appropriate formulation of the coalition and clubs model is to express it in terms of households. Households are faced with alternative sizes of city (either an historically determined city-size distribution that reflects past rates of natural increase and the cumulative effects of previous migration decisions, or a full spectrum of hypothetical city sizes). Each city can be treated as a coalition of households offering a certain mix of costs and benefits. Among these are included levels of public goods and services and the associated tax-expenditure mix, so that the analysis may be treated as an extension of Tiebout's (1956) community-preference model in which people migrate to associate with others and to share their preferences for public goods and tax-service levels. This extension is all the more appropriate now that Tiebout's analysis has been amplified to include intraurban residential-location decisions where households choose where to live within a large metropolitan area (separate city, school district, etc) according to the same kind of criteria (Oates, 1969; Ellickson, 1971). However, the analysis need not be confined to public goods and services as they are narrowly defined. Other consumption externalities—leisure and cultural amenities, clean air, climate—may be included among the benefits, as well as more obvious considerations such as job opportunities, and the less obvious such as life-style preferences. It is clear that this model relies heavily on the assumption that households have heterogeneous tastes and preferences, and living in cities of different size is an important manifestation of this heterogeneity. Accordingly, if this model were to be used for optimum geography it would be inconsistent to use the simpler, homogeneous-household intraurban models. On the contrary, it would appear sensible to use those residential-location models that take adequate account of the spatial distribution of public services, clean air, and other externalities. Furthermore it may be possible to include nonsubstitutable preferences for 'lifestyle-area' mixes (for example, outer suburbs versus close to downtown) or housing types (single-family detached dwelling versus high-rise apartment). Conceptually, this implies that the conditions of locational equilibrium and equal utility apply only within household strata. One consequence of this type of approach may be a simpler explanation of residential segregation patterns.

If we return now to the interurban case, each household tries to join the coalition (city), membership of which maximizes its own net benefits. The obvious way of formalizing this and of reconciling the interurban with the intraurban location is to introduce city size as an argument of the utility function, that is

$$\text{maximize } U = U(c_{ri}, s_{ri}, N_i) \,, \tag{10.13}$$

where N_i is a city of size i, and stands as a surrogate for all the net

An optimum geography 143

benefits (benefits minus costs) associated with living in a city of that size,

subject to $y_i - t_{ji} = v_i c_{ri} + p_{ri} s_{ri} + t_{ri}$, (10.14)

where t_{ji} are the migration costs between a city of size N_j and a city of size N_i. If the household is already living in a city of the preferred size then $t_{ji} = 0$. Living costs (the price of composite consumption, land values, and transport costs) vary with city size. However, apart from the fact that incomes and costs (and hence consumption) are city-size-specific, the conditions for intraurban locational equilibrium remain unaffected.

The process by which households choose their favoured coalition results in a distribution of city sizes that is stable. If so many households migrate into a particular city that the city outgrows the acceptable range of variation in its coalition size for many of the 'members', the latter can subsequently move to smaller cities without disturbing the hierarchy. The city-size distribution itself may remain highly stable even if individual cities jump ranks quite frequently. The problem, however, is whether or not this model generates a distribution of city sizes of the kind observed empirically. In particular, what is required is an urban hierarchy where the number of cities in a given size class declines as cities increase in size. Whether this is likely depends upon the number of households seeking locations in cities of a particular size. For example, it is possible to obtain a large number of very small cities only if a great many households maximize their net benefits in such cities. One possible assumption leading to the desired result is what Parr (1970) calls the equal hierarchical population (EHP) assumption. The number of cities of size N_i is determined by the total population, P_i, wanting to live in cities of that size. If the population wanting to join each size class is equal ($P_1 = P_2 = P_3 = ... = P_i = \bar{P}$), the city size distribution looks reasonable. The number of cities in each size class ($K_1, K_2, ..., K_i$) will be equal to $\bar{P}/N_1, \bar{P}/N_2, ..., \bar{P}/N_i$, respectively. As required, the number of cities increases as city size declines. Although the reasonableness of the EHP assumption is sensitive to how city-size classes are defined, it is compatible with the hypothesis that preferences for living in cities of different sizes are randomly distributed. This is also equivalent to a version of the rank-size rule where the rank refers to size class rather than to an individual city. Moreover, it is consistent with a simple form of entropy (maximum uncertainty) model in which each size class (rather than a specific city) is equally likely to be chosen by an individual household *a priori*. Although these points do not amount to an 'explanation' of household preferences for different city sizes, they at least show that the one additional assumption needed to obtain a plausible city-size distribution has very similar counterparts in quite different strands of city-size distribution theory.

Swanson et al. (1974) have recently attempted an interesting reconciliation between optimal city size and theories of the distribution of city size. Their model is based upon plant location decisions, with firms choosing locations according to relative capital, labour, and public service costs. Some interesting discoveries were made, such as the interdependence between changing technology and the city-size distribution; for example, increasing capital intensity of production is associated with an increase in average city size. Nevertheless their study falls short of a comprehensive general equilibrium model because it is *assumed* that the cost of capital is inversely related to city size and the labour costs directly related, whereas the function relating cost of public services to city size is assumed to be U-shaped. Moreover, and more critical from the optimum-geography viewpoint, their model is nonspatial, since it ignores land as an input and the role of land rent in intraurban allocation. However, they do recognize that land rent is an equilibrating variable in interurban allocation, but that in the long run the benefits of city size accrue only to landowners (cf Edel, 1972). At bottom, this study is a variant of the theory of the distribution of city sizes that places negligible emphasis on what happens within cities. It cannot, therefore, claim to be an optimum-geography model.

Hierarchy models
Some of the more satisfactory economic theories of the distribution of city sizes are based upon the production sector, and hence are difficult to reconcile with the NUE models that focus on the household sector. A good example is Tinbergen's hierarchy model (1968)[24]. He assumes a regularly-shaped closed economy evenly covered with farms, except in cities. There are H industries ($h = 0, 1, 2, ..., H$), where h is the industry rank ($h = 0$ in agriculture). Demand for product h is satisfied by n_h firms (assumed to be of optimal size), and is equal to $a_h Y$, where Y is the national income and a_h is the demand ratio for h. It is assumed that there is only one firm in the highest-ranked industry ($n_H = 1$), and that the number of firms in each industry varies with rank ($n_1 > n_2 > n_3 > ... > n_H$). All income is spent, so that $\sum_h a_h = 1$. Rural income, Y_o, is $a_o Y$.

To obtain predictions about the size distribution of cities, Tinbergen makes additional assumptions. There are M orders of cities ($m = 1, ..., M$).

---

[24] Davis and Swanson (1972) have developed an interesting theory of city-size distribution that makes the rate of growth of the labour supply in a city a function of capital accumulation, real wages, and technical progress. By arguing that technical progress contains a random component, they are able to show that the framework predicts a lognormal city-size distribution. Although their model has the considerable virtue of introducing dynamics into the theory of city-size distribution, it is not helpful for optimum-geography analysis for two main reasons: it is a macroeconomic model, and the complementary intraurban assumption would be either spacelessness or at best exogenously-located CBD production; it contains no equilibrium growth rate, so we have no idea as to what its equilibrium conditions, if any, might be.

In any city of rank m only the industries appear for which $h < m$. The industry of rank h in city m satisfies local demand, and exports downward to smaller cities (each city m exports the same amount of commodity h). He assumes that there is always *only* one firm of the highest rank in each centre. From these assumptions he derives

$$Y = Y_o + \sum_m \sum_h Y_h^m = \frac{a_o Y}{1 - \sum_h a_h} \, , \qquad (h \neq 0) \, , \tag{10.15}$$

and

$$K^m = n_h \frac{a_o}{1 - \sum_h a_h} \, , \qquad (h \neq 0) \, , \tag{10.16}$$

where K^m is the number of cities of rank m.

These two equations determine the size distribution of cities.

Whether this size distribution represents an optimum is unclear. In the context of the model, optimality means the minimization of total transport costs for all goods entering interurban trade. Tinbergen showed that in the simple case where $H = 2$, the model generates an optimum if transport costs depend on the type of good transported rather than on distance. If transport costs depend on distance, additional assumptions are needed. It is possible to generalize the model by introducing foreign trade, natural resources and other uniquely located industries, intermediate goods, and less restrictive interurban trade flow patterns, but the implications for optimality are unclear. An empirical test of the model making use of French data and assuming seven ranks, a city nesting ratio of four, and a five-fold increase in the number of firms with each downward step in rank, yielded a city-size distribution very different from the actual one. In particular the model predicted a smaller size for Paris, but then predicted too much concentration in other high-ranking cities. It is uncertain "whether the deviation means nonoptimality in the actual distribution or lack of realism in the theoretical one" (Tinbergen, 1968, page 68). Since the primate characteristics of the French national urban hierarchy are well known, the former explanation is reasonably plausible.

How easily this model might be made compatible with an intraurban model is not obvious. Tinbergen implies that extended versions of his model would introduce space in the form of transport cost-distance functions, even though the model becomes more complicated as a result. Similarly, a compatible intraurban model would need to take account of space and location. A basic implication of the model is that within each city there is a hierarchy of industries in which the number of firms in each industry increases with descent of the hierarchy. If these industries supplied local markets (that is, within the city), their location might be predicted via an intraurban variant of the Löschian central-place model. This assumption holds for every industry except that of the highest rank in each city. Nevertheless there are several dangers in moving in this direction. This particular assumption is one of the most dubious in the Tinbergen model. For instance, there may be industrial specialization which may break up the correspondence between export-industry rank and centre rank. It is quite common for large firms located in small cities to export *up* the hierarchy. The empirical evidence on intraurban nonresidential location suggests the relevance of the Löschian model only for certain types of service sectors. Most important of all, the highest-ranked industry may exert—via its stimulus to agglomeration economies—a locational pull on all other industries. Yet where the highest-ranked industry locates cannot be predicted without specifying the industry. In some cases, it might be the CBD; in others, the optimal location may be in the suburban periphery. To resolve these issues requires much more research on the theory of *intraurban* nonresidential location than has been undertaken hitherto. The fact that intraurban and interurban models have been developed in isolation from each other is the chief obstacle in the way of developing a general equilibrium optimum-geography model. Theorists in each branch of theory have ignored the repercussions of their analysis on the other.

Space and optimum geography
If the distinctive feature of NUE is its analysis of how rents and densities vary over space, then the optimum-geography models should also deal explicitly with space. Economic geographers have obviously devoted attention to this, but predominantly in terms of empirical analysis or in the derivation of statistical rules. Common examples include Stewart's hypothesis (1958) that interurban distance is proportional to the multiple of city populations ($d_{ij} = N_i N_j/M$ where M is a constant) and the Christaller-Lösch principle that $d_i = d_{i-1} K^{0.5}$, where K is equal to the number of cities of rank $(i-1)$ supplied by a city of rank i. The value of K depends on the nesting pattern in an urban hierarchy. Curry (1967) preferred the special case of $d_i = d_{i-1}^{1.5}$.

The theory of interurban distances can be deduced from market-area analysis (Lösch, 1954), but it does not hold up very well in empirical tests—apart from the general point that the distance between cities of similar size tends to increase with city size. The reasons are: the inaccuracy of the underlying assumption of uniform population densities; the fact that most countries have been developed initially by external sources so that the primate city tends to be located peripherally rather than centrally; and the influence of national population size and of the geographical area on average interurban distance. As a result, the symmetry of the Christaller-Lösch homogeneous plain is neither necessary nor helpful.

One possibility is to assume a landscape of uniform rural population densities and rural rents, relieved by rental and density peaks of various heights, to reflect the city size distribution. However, even these assumptions may be relaxed. Provided that the boundary conditions between urban and rural areas are satisfied, it is permissible to allow some heterogeneity in the rural landscape. For example, if the principle of differential fertility and/or productivity variations over space are permitted, it is acceptable for rural rents to vary spatially. This is consistent with Schultz's observation (1953) that the efficiency of agriculture tends to be higher closer to cities. It is also plausible that spatial rent differentials will disturb the pattern of uniform density, though in what way is unclear since the theory of rural population distribution has been grossly neglected by spatial economists.

If the regular spacing of traditional central-place models is not necessarily a characteristic of an optimum geography, how will the many small cities be located relative to the few large metropolises? Much depends on the assumptions. If the features of specialization among cities that are derived from the standard hierarchy model are assumed, with exporting only down the hierarchy, the spatial distribution is likely to be more regular than under alternative assumptions. However, if smaller cities compete in some sectors with large cities, a different pattern may result. For instance, von Böventer (1970) suggested that the optimal distance for a city from the nearest larger city is where the sum of agglomeration and hinterland effects are maximized. The agglomeration effect is the spatial spillover benefits of agglomeration economies obtained from being close to a larger city; the hinterland effect is the shelter from the competition of other cities received by being far away from them. The worst location is at an intermediate distance where agglomeration effects have dwindled to almost zero, but the monopoly shelter is not yet assured.

Agglomeration and hinterland effects will vary over time. Since individual cities cannot change their location, and most space economies are not built simultaneously from scratch but reflect a historical legacy of economic and social development, the dynamics of national and regional urban hierarchies are primarily represented by the differential growth of *existing fixed* cities rather than by the growth of new cities. The distortions owed to the past (including the role of *locational constants*; see Richardson, 1973c) mean that the optimal spatial distribution at any point in time is different from the normative spatial pattern that might result if an empty country was planned from the beginning. However, provided that the conditions for nonspatial maximization are satisfied, the differences in welfare between the positive and the normative spatial patterns should be relatively minor. Much of the argument on this point has revolved around the controversy about whether the primate cities in a market economy grow too big because negative net externalities are not taken into account in migration and business decisions. However, that the net externalities of big cities are negative is not proven (Richardson, 1973b); in general, the external benefits of city size have received considerably less attention than the external costs. Perhaps what is more important, to the extent that externalities are internalized in the land and labour markets, is that the market itself applies a degree of corrective action. On this view big cities offer higher wages and lower land rents than in the absence of external diseconomies because these are needed to attract migrants (households especially, but possibly some businesses too) from more desirable locations in smaller cities. The factor price differentials are a condition of equilibrium rather than a symptom of disequilibrium. To accept the alternative view is to assume that migrants are so ignorant and irrational that they believe that their welfare depends on money income alone. If this were so, migration flows to large cities would have been much heavier than they have been and income differences would have been much narrower.

An interesting paper by Weiss (1961) gave some attention to the questions of linking the intraurban density gradient with the rank-size rule and the interurban spacing relation ($d_{ij} = N_i N_j / M$). Although his analysis was statistical (using United States Census data) and heuristic, and lacked a theoretical base, he was able to use these relationships to shed light on such questions as the minimum costs of servicing a national population, and the minimum land area needed for containing the population. The latter analysis showed that one-quarter of the US population lived on a tiny area of about three thousand square miles. This implies, for instance, that Mirrlees's assumption that the land area constraint is not binding is reasonable. Despite the absence of theory in Weiss's study, it does suggest that there are insights to be gained from an explicit combination of intraurban and interurban concepts and models. It is desirable that future work on the theory of optimum geography should include research on these lines.

Conclusion
It would not be appropriate to describe this chapter as a review of the theory of optimum geography. Such a literature does not exist in sufficient quantity to merit a review. There is a voluminous body of research on city-size distribution theory that could be integrated with some of the NUE intraurban models. However, this has been largely ignored here on the grounds that such an integration would be a clumsy graft between two quite different branches of theory. Mirrlees (1972) made a noble attempt at achieving a 'grand design', but his assumptions were unnecessarily restrictive, his identical production function treatment could and should have been avoided, and the neglect of interurban space was a serious flaw.

Probably Mirrlees was severely handicapped by starting with the intraurban model and working outwards to the economy at large (of course, his choice was a conscious one since his preoccupation was with the distribution of utility *within* cities). A more logical approach would be to begin with the interurban theory and then descend to the intraurban level. One reason is that the main problem is to explain the size and spatial distribution of cities that both satisfies equilibrium and/or optimality conditions and corresponds, at least roughly, with reality. It is not difficult to generate any particular city size with a standard intraurban NUE model, because urban rent is an effective adjustment variable via its impact on the urban boundary, the rent gradient, and hence on densities. Also, intraurban factors may help to introduce irregularities into the city-size distribution predicted by some models. For instance, many central-place hierarchy models have the drawback that they generate equal-size cities within each hierarchy rank. A less rigid urban hierarchy would be more satisfactory if it could be shown to be compatible with equilibrium. Negative externalities, such as congestion or pollution, may help to explain why some cities do not grow as much as others (even within the same rank). Differences in commuting costs between cities could have the same effect.

There are several other guidelines that may be helpful in the development of an optimum-geography model. First, since the distinctive feature of NUE intraurban models is their spatial analysis (though mainly dealing with one-dimensional space) the interurban component of the model should also include the spatial aspects of the distribution of cities. This requirement rules out, at least in the absence of drastic modification, most existing city-size distribution theories as irrelevant, since almost all of them are spaceless. Second, it is easier, as well as very sensible to introduce multiple sectors and economies of scale to explain the city-size distribution (see Beckmann, 1958; Tinbergen, 1968; Zipf, 1949). Consequently, the intraurban component ought to include these characteristics too. Dixit (1973) analyzed the question of economies of scale in a factory town reasonably comprehensively, but no NUE modellers (apart from those developing discrete models, such as Mills, 1972b, and Hartwick and Hartwick, 1974) have been sufficiently specific about the production sector to introduce many commodities. Of course, plants producing different commodities might have different locational requirements, and the assumption of all nonresidential establishments located in the CBD is much shakier. Third, and not unconnected with both previous points, the transportation sector needs more attention than it has been given, both between cities and within cities. In the latter case this means that freight transportation should be analyzed as well as commuter transportation. Fourth, first attempts at an optimum-geography model should not be overambitious. For instance, some variables will need to be determined exogenously for some time to come. A good example is the location of some cities in the economy, since building up an optimum geography from first principles (for example, on a Löschian homogeneous plain) usually leads to an unrealistic spatial pattern. Fifth, a useful assumption is to allow for heterogeneity of tastes. If people differ in their preferences about the size of city in which they wish to live, it places much less strain on the equal utility condition.

Two residential location models

1 *The possibility of positive rent gradients*
A claimed virtue of NUE models is that they predict plausible spatial distributions; in particular, downward-sloping, rent-distance functions and density gradients. In fact there is some empirical evidence in favour of a positive rent-distance function (Richardson *et al.*, 1974a; Wilkinson, 1972). and this may be consistent with theories of residential location that stress neighbourhood amenities and environmental preferences rather than those that depend on a trade-off between accessibility and space. Once the competition of nonresidential establishments is assumed away, as in NUE models, the case for the negative rental function becomes much weaker. A positive rent-distance function is ruled out in these models (and in standard urban theory) because it would then be impossible to achieve locational equilibrium. If rents increased with distance, households could move towards the city centre in order to cut down on transport costs, but the increased demand for limited space would then force up the rents.

Given a Cobb-Douglas logarithmic utility function

$$U(c, s) = g \log c + h \log s ,$$

and a budget constraint

$$c + p_r s_r + t_r = y ,$$

where c, s, p, t, y, and r have their usual meanings, and g and h are coefficients; first-order conditions for utility maximization are, as noted earlier (page 43),

$$\frac{g}{c} = \frac{h}{p_r s_r} ,$$

and

$$p'_r s_r + t'_r = 0 .$$

The latter implies that in equilibrium a household at r consumes space in such a way that any marginal change in housing expenditure would exactly offset a change in transport cost. Since transport costs increase with distance, equilibrium demands that the rental function be negative. Although many of the urban models are more complicated than this, the negative rental gradient remains a core finding.

Externality rent
The objective here is to show that a positive rental gradient is feasible as a result of introducing an externality component into the determination of urban rent, and that such a gradient is consistent with locational equilibrium.

The idea of externalities as an element in rent is discussed by Mirrlees (1972), and implied by Yamada (1972), but its significance is missed because they both start out by *assuming* that the rental gradient is negative.

Whereas the standard model assumes that location rents are the sole component of urban rent, the argument here is that there is another element in rent, called *externality rent*. Externality rent arises from the fact that specific sites confer advantages in addition to their accessibility to the CBD. These advantages could arise from a myriad of influences, but in residential-location models the most common are neighbourhood externalities, in the form of area amenities and pleasant living environments, and all the social features associated with them. In a theoretical model it is convenient to use low population density as a surrogate for these environmental externalities (Mirrlees, 1972; Wabe, 1971).

It is plausible, though not strictly necessary, to allow the preference for low-density neighbourhoods to be a direct function of income. (Examples that employ both identical and unequal incomes are discussed below.) The rich may be more willing to pay high externality rents because they value the environmental amenities and qualities associated with low densities. Given the preference of high-income groups for low densities, and their ability to pay, income may be expected to increase with distance as density falls. Also, externality rent increases with outward movement and absorbs a larger share of total rent. Transport costs increase and hence location rents decline with distance. What happens to the overall rent–distance function depends upon whether rising externality rent compensates for falling location rent.

Locational equilibrium
In the standard model with constant incomes a positive rent gradient is impossible because rents are purely location rents, arising from savings in transport costs. Abstracting from a rural rent floor, urban rents decline to zero when all income net of consumption is absorbed by transport costs. When multiple-income groups are permitted this constraint becomes less severe. However, in this case the higher housing expenditures made possible by higher incomes are spent on more space *at lower unit rents*. The rich live further out because a negative density gradient is necessary for overall equilibrium. Since the urban area is an increasing function of distance (πr^2), density must fall with distance for the supply and demand for land to be in equilibrium at each location.

In the model of the multiple-income group, a positive rent gradient is a short-lived possibility, but it cannot be maintained because it is incompatible with equilibrium. Since everyone desires accessibility (because it economizes on transport costs), higher rents at greater distances would encourage households to relocate by moving closer to the city centre. The supply of land at each distance is fixed, and these attempts to relocate would add to demand at close-in locations and subtract from demand further out.

Thus market conditions would increase close-in rents and reduce rents at greater distances. Hence a condition for locational and market equilibrium is that rents must decline with distance.

These arguments follow from the assumption that rent is pure location rent, where location is defined narrowly in terms of distance from the CBD. The way out of the dilemma is to introduce some other element in rent determination, which tends to become larger with increasing distance. The solution suggested here is to develop the concept of externality rent, which is negatively associated with density. Market rent contains a component that expresses a preference for low-density environments. Although this is an externality in the sense that density is determined by the location decisions of others and that the quality of neighbourhoods is external to the attributes of the specific site, it is nevertheless reflected in the price of land. The importance of low-density preferences in this model is that it acts as a constraint on inward relocation even if rents are lower nearer the city centre. Such relocation might bring about a benefit in terms of lower transport costs but it would also involve a cost in terms of higher densities. If the environmental preferences are strong enough relative to the benefits of accessibility, a positive rental gradient is compatible with equilibrium. In this case moving closer to the city would save transport costs and permit lower unit rents, but there is no net welfare gain (even if no other household attempts to relocate) because the monetary savings are offset by the disutility of having to tolerate higher densities.

Theory

The preference for low densities has to be explicitly introduced as an element in household utility, and rent payments must then include externality rent. However, if the preference for low densities is internalized in the land market, the inverse relationship between total rent and transport expenditures may be broken.

Case 1: equal incomes

If we assume a monocentric city of identical households, the concept of externality rent is most easily handled by introducing neighbourhood density as an argument of the utility function. Thus

$$U = U(c, s, \bar{s}) , \tag{11.1}$$

where
- c is composite consumption,
- s is the amount of land consumed, and
- $\bar{s}$ is the average plot size in the neighbourhood (or the reciprocal of density).

To obtain equilibrium each household maximizes its utility, thus

$$\text{maximize } U = U(c, s, \bar{s}) , \tag{11.2}$$

Two residential location models

subject to the budget constraint

$$vc + p_r s_r + t_r = y \,, \tag{11.3}$$

where
v is the price of composite consumption good,
p_r is the rent at distance r,
t_r are transport costs, and
y is income,

and subject to

$$s_r > 0 \,. \tag{11.4}$$

Allowing p_r^x to represent externality rent, or the shadow price of low-density environments, the following first-order maximization conditions may be obtained:

$$u_c - \lambda v = 0 \,, \tag{11.5a}$$

$$u_s - \lambda p = 0 \,, \tag{11.5b}$$

$$u_{\bar{s}} - \lambda p^x = 0 \,, \tag{11.5c}$$

$$\lambda(y - vc - ps - t) = 0 \,, \tag{11.5d}$$

where the us are the respective marginal utilities, and λ, the Lagrangian multiplier, is the marginal utility of income.

The equilibrium conditions are:

$$\lambda = \frac{u_c}{v} = \frac{u_s}{p} = \frac{u_{\bar{s}}}{p^x} \,, \tag{11.6}$$

$$s_r = \bar{s}_r \,, \tag{11.7}$$

and

$$p_r' s_r + t_r' - p_r^{x'} \bar{s}_r = 0 \,. \tag{11.8}$$

Rearranging equation (11.8), we obtain

$$p_r' = \frac{p_r^{x'} \bar{s}_r - t_r'}{s_r} \,. \tag{11.8a}$$

Since $s_r > 0$,

$$p_r' \gtreqless 0 \,, \quad \text{according to whether} \quad p_r^{x'} \bar{s}_r \gtreqless t_r' \,. \tag{11.9}$$

Thus the rental gradient is positive if the marginal increase in the externality rent of the site occupied ($s_r = \bar{s}_r$) with outward movement exceeds the marginal increase in travel costs. Further light may be shed on the conditions for a positive rental gradient by substituting $(u_{\bar{s}}/u_s)p_r'$ for $p_r^{x'}$ from equation (11.6), and s_r for $\bar{s}_r$ from equation (11.7) into equation (11.8).

As a result

$$p'_r = \frac{-t'_r}{(1 - u_{\bar{s}}/u_s)s_r} \ . \qquad (11.10)$$

The condition for $p'_r > 0$ is that

$$\frac{u_{\bar{s}}}{u_s} > 1 \ , \qquad (11.11)$$

that is, the marginal utility of low density must be greater than the marginal utility of the individual site size. This implies that $p_r^{x'}$ must be larger than p'_r, that is, the share of externality rent in total rent must increase. This condition illustrates very clearly that more household space is not equivalent to lower areal density. The rent a household is willing to pay for a site in a particular neighbourhood will contain an externally-determined component that accrues from the site of surrounding sites rather than from the individual site per se. This component is the externality rent.

Case 2: unequal incomes

If city residents earn different incomes, the above model needs to be modified in at least two ways: externality rent may vary with income (the demand for low-density environments is income-elastic); and travel time, the value of which varies with income, should be introduced explicitly. One possible approach is to make the components of time arguments of the utility function. Further, assume that income consists solely of wage income. Thus the problem is

$$\text{maximize } U = U(c, s, \bar{s}, \theta^\ell, \theta^c, \theta^w) \ , \qquad (11.12)$$

subject to

$$vc_r + p_r s_r + t_r = y_r \ , \qquad (11.13)$$

$$y_r = w_r \theta^w \ , \qquad (11.14)$$

$$\Theta = \theta^w + \theta^c + \theta^\ell \ , \qquad (11.15)$$

$$s_r > 0 \ , \qquad (11.4)$$

where
w_r is the unit wage at distance r,
Θ is the total time available,
θ^ℓ is leisure time,
θ^w is working time, and
θ^c is travel time.

With this model, the equilibrium conditions are

$$\lambda = \frac{u_c}{v} = \frac{u_s}{p} = \frac{u_{\bar{s}}}{p^x} = \frac{u_{\theta^w}}{w} = \frac{u_{\theta^c}}{v_{\theta^c}} = \frac{u_{\theta^\ell}}{v_{\theta^\ell}} \ , \qquad (11.16)$$

Two residential location models

where
v_{θ^c} is the value of travel time, and
v_{θ^ϱ} is the value of leisure time.

$$\lambda(\theta^{w'}w_r + \theta^{c'}v_{\theta^c_r} + \theta^{\varrho'}v_{\theta^\varrho_r}) = 0, \qquad (11.17)$$

$$s_r = \bar{s}_r, \qquad (11.7)$$

$$p'_r s_r + t'_r + v_{\theta^c_r}\theta^{c'}_r - p^{x'}_r \bar{s}_r = 0. \qquad (11.18)$$

Equation (11.16) expresses the first-order conditions for utility maximization whereas equation (11.17) illustrates the equilibrium conditions for the allocation of time. The locational-equilibrium condition, equation (11.18), is adjusted to include the effect of travel time. Accordingly the condition for $p'_r > 0$ is similar to equation (11.9) modified to allow for travel time, thus

$$p'_r \gtreqless 0, \qquad \text{as} \qquad p^{x'}_r \bar{s}_r \gtreqless t'_r + v_{\theta^c_r}\theta^{c'}_r, \qquad (11.19)$$

and

$$p'_r = \frac{-(t'_r + v_{\theta^c_r}\theta^{c'}_r)}{(1 - u_{\bar{s}}/u_s)s_r}. \qquad (11.20)$$

From equation (11.19), the value of the externality associated with the site must now be higher to ensure that $p'_r > 0$, but, from equation (11.20), the condition for a positive rent gradient, that $u_{\bar{s}} > u_s$, remains unchanged.

The interesting question in the unequal-income model is how does income vary with distance so as to satisfy $p'_r > 0$? This can be explored easily if the change in externality rent is assumed to be functionally related to the change income, that is

$$p^{x'}_r = by'_r. \qquad (11.21)$$

Two cases may be considered. The simplest is where the hours of work are fixed ($\theta^w = \bar{\theta}^w$). In this case the time constraint (11.15) reduces to

$$\theta^\varrho = J - \theta^c, \qquad (11.22)$$

where $J = (\Theta - \bar{\theta}^w)$.
Substituting equation (11.21) into constraint (11.19) and rearranging, we obtain

$$y'_r \gtreqless 0 \qquad \text{as} \qquad \frac{t'_r + v_{\theta^c_r}\theta^{c'}_r}{b\bar{s}_r} \gtreqless 0. \qquad (11.23)$$

The *a priori* expectation that $y'_r > 0$ holds if $b > 0$ [the numerator in the inequality (11.23) is positive since the money and time costs of travel both increase with r]. Thus the rich will live further out—for the case where the rent gradient is positive—if the preference for low-density environments (reflected in a willingness to pay externality rent) is positively related to income.

The case where hours of work are variable is a little more complicated because time must now be allocated optimally between all three of its uses. The value of additional travel time with increasing distance is equal to the value of lost income (owing to reduced working time) plus the value of sacrificed leisure. With the aid of equations (11.17) and (11.21), the constraint (11.19) may be rewritten as

$$p'_r \gtreqless 0 \quad \text{as} \quad by'_r \bar{s}_r \gtreqless t'_r + w_r \theta^{w'} + v_{\theta_r^\varrho} \theta^{\varrho'} \,, \tag{11.24}$$

and since

$$y'_r = (\theta^w + \theta^{w'})(w_r + w'_r) - \theta^w w_r \,, \tag{11.25}$$

or

$$y'_r = w_r \theta^{w'} + w'_r \theta^w + w'_r \theta^{w'} \,, \tag{11.25a}$$

constraint (11.24) may be rewritten as

$$p'_r \gtreqless 0 \quad \text{as} \quad b\bar{s}_r w'_r(\theta^w + \theta^{w'}) \gtreqless t'_r + v_{\theta_r^\varrho} \theta^{\varrho'} \,. \tag{11.26}$$

It is clear that $y'_r > 0$ if

$$w'_r(\theta^w + \theta^{w'}) > -w_r \theta^{w'} \,, \tag{11.27}$$

that is, if the increase in wages with outward movement (noting that $\theta^{w'} < 0$) is greater than the lost income owing to shorter working hours (the result of increased travel time).

Rearrangement of constraint (11.26) enables inferences to be drawn about the wage gradient, w'_r, neighbourhood site size, $\bar{s}_r$, and the marginal externality-rent/income ratio, b, for determining the sign of the rental gradient. The reader is left to work out the implications for himself. However, it may be noted that since $(t'_r + v_{\theta_r^\varrho} \theta^{\varrho'}) > 0$, a necessary but not sufficient condition for $p'_r > 0$ is that the product of the variables on the left-hand side of constraint (11.26) should be positive. Since $\bar{s}_r > 0$ [from constraint (11.4) and the equilibrium condition (11.7)] and $(\theta^w + \theta^{w'}) > 0$ even though $\theta^{w'} < 0$—unless the householder does not work at all, in which case he neither earns income nor travels to work; much depends on the values of b and w'_r. If the wage gradient is negative ($w'_r < 0$), p'_r must also be negative, unless $b < 0$. It is more plausible that $b > 0$, and in this case w'_r must also be greater than zero for a positive rental gradient to be a possibility. Even if $w'_r > 0$, $p'_r > 0$ only if the parameters in constraint (11.26) ensure that its left-hand side is greater than the right. Also, it is clear from inequality (11.27) that $w'_r > 0$ is a necessary but not sufficient condition for $y'_r > 0$. The increase of wages with distance must be sufficiently strong to offset shorter working hours resulting from increased travel time.

Conclusion
This analysis has shown that the virtually unanimous prediction of a negative rental gradient is the result of an unnecessarily narrow treatment of urban land rent as location rent, determined by the familiar access–space trade-offs of residential location theory. Both *a priori* reasoning and some empirical studies suggest that the choice of a residential site involves consideration of externalities in the form of good and bad features of the neighbourhood where the household locates, such as neighbourhood amenities and characteristics of environmental quality. The density of the neighbourhood is an acceptable surrogate for these externalities, where low density may be equated with a good environment. These external attributes have a marginal shadow price, defined here by the expression $p_r^{x'} = (u_{\bar{s}}/u_s)p_r'$. This modification to the standard urban model makes a positive rental gradient entirely plausible and compatible with locational equilibrium. This result is obtained without introducing subcentres which may lend to pseudopositive sections on the rent gradients. Although the introduction of nonresidential establishments, some of which will desire a central location, into the model will tend to make the rental gradient negative close to the CBD, even a positive section on some part of the rental gradient is disturbing both to traditional urban theory and to that of the 'new urban economics'. The conditions for $p_r' > 0$ obtained here suggest that a positive rental gradient is much more than a special case. If this is so, the implications for the direction of urban economic theory are far-reaching.

2 *Discontinuous densities, urban spatial structure, and growth*
Introduction
Many problems in regional and urban analysis have been investigated with the aid of analogies derived from the physical sciences. They include the gravity model and potential analysis in a wide variety of applications (migration, travel behaviour, market-attraction models, etc), electric circuits to simulate spatial price analysis, thermodynamics in entropy quasi-equilibrium models, relativity theory in time–space analysis, energy transfers and the natural transport rate, the transmission of radio waves and diffusion analysis, and allometric growth in the theory of city-size distributions. In only one or two cases have these analogues been made consistent with economics. Although they have been useful and have added new insights, the analogies have frequently been diverting because of the user's close adherence to the framework of the original science.

The kinetic theory of gases is used here as a starting point for the analysis of certain elementary principles of urban spatial structure and growth. Although the analogy involved is fairly close—the effect of temperature in increasing, and of pressure in reducing, gas volume, compared with the influence of income and rents on population densities—the analysis quickly moves away from the physical sciences. The main purpose of the scientific analogy is to provide a useful mathematical equation (van der Waals equation). The structure of this equation is important because it can be used to generate changes in densities that could not be predicted by standard urban-density models, whether theoretical (neoclassical utility maximization; for example, Solow, 1972) or empirical (the negative-exponential density gradient; see Clark, 1951; Mills, 1970; 1972a; Muth, 1969; Niedercorn, 1971).

Although the framework is strikingly simple by the standards of present-day urban models, and despite its physical origins, the theory corresponds to plausible economic principles. First, the main parameters—income, rents, and density—are the key economic variables of modern urban economics. Second, the main elements of the model could be explained in terms of neoclassical analysis, though for several reasons, but above all because of the nonmarginalist results, it may be preferable to describe the functions in behavioural terms. Third, the intertemporal version of the model enables change to be analyzed over time, and this dynamic approach contrasts sharply with the predominantly static models developed hitherto. In particular the model can allow for variations in income, whereas in many urban models the income level is given. Last, and most important of all, this model generates an important phenomenon in real-world cities which cannot be obtained with the standard models, namely discontinuous shifts in density (either over space with variations in distance from the CBD or in the city as a whole over time).

Density, income, and rents

The simple functional relationship

$$s^{-1} = f(y, p) , \qquad (11.28)$$

where
s^{-1} is the density (households per acre),
y is the income (per household), and
p is the rent per acre (which may be capitalized as land value),
offers interesting prospects for analysis in urban economics. It is applicable to either spatial or intertemporal analysis. In the former case, density at a given distance becomes a function of the income level of residents there and of the local rent level. Since rents normally fall and income rises with distance this is quite consistent with the density-gradient concept. In intertemporal analysis, if the variables are treated as city-wide averages, the framework enables changes in city densities to be determined by changes over time in the level of urban income and in the aggregate rent structure[25].

[25] Transportation changes are implicitly subsumed in rents and incomes. It is easy to introduce transportation costs, t, by transforming the income variable into net income after transport costs, that is $(y-t)$. A reduction in transport costs generally increases income and flattens the slope of the rental and density gradients. In a more comprehensive model, it may be useful to treat it as a separate parameter (as in most of the NUE models discussed in this book), but most of its impacts work indirectly through rents and income effects.

This advantage is important in view of the scarcity of dynamic models. If the spatial structure of the city can be expressed in terms of its densities, the model presents a valuable, if highly aggregated, framework for dynamic analysis. Since s^{-1}, as a measure of average density, can be translated into a density gradient, and additional equations can be introduced to determine changes in income and rents, the model can be expanded into an integrated model of urban growth and structure. The prediction of a discontinuity in urban densities over time is very important for providing theoretical support for relatively sudden shifts in the pace of decentralization.

The basic model
The new urban economists' love of calculus in general, and calculus of variations (Pontryagin's principle) in particular, has restricted the scope of their analysis to continuous models. Because of subcentring, topographical constraints, residential segregation, heterogeneous preferences among income groups, the limitations of the urban transport network, and many other characteristics of modern cities, there is much in favour of models capable of dealing with or generating discontinuities. Apart from zonal disaggregation of the city (discrete models; see Mills and MacKinnon, 1973), a quite different approach has been suggested by a British mathematician, Amson (1974). His approach uses an analogy derived from physics, in particular from the kinetic theory of gases. The purpose of this paper is to develop Amson's analysis and to translate it into economic terms.

The starting point is the 'basic model', a simple urban relationship analogous to the 'ideal gas' equation[26]:

$$p = Kys^{-1}, \qquad (11.29a)$$

and its corresponding density function

$$s^{-1} = \frac{p}{Ky}, \qquad (11.29b)$$

where K is a positive constant. This relationship implies at least three important urban economic relationships:

---

[26] The ideal gas equation is $pV = RT$, where p is the pressure, V is the volume, T is the temperature, and R is a constant. A special case where temperature is constant is Boyle's Law ($pV = K$).

The corresponding equation to the discontinuous density model is van der Waals equation, merely one of at least fifty-six ways of modifying the ideal gas equation. The van der Waals equation is

$$\left(p + \frac{a}{V^2}\right)(V - b) = RT,$$

where a and b are constants. The terms a/V^2 and $-b$ are corrective terms for pressure and volume respectively. This equation is identical to equation (11.32b) discussed below where, $p = p$, $V = 1/s^{-1}$, $T = y$, $a = \alpha$, $b = \beta$, and $R = K$.

(1) Rent is a direct function of income and density. Higher incomes boost housing demand, whereas higher densities reflect, *inter alia*, intensified competition for limited urban space.
(2) Housing expenditure per household is a function of household income ($ps = Ky$). Equation (11.29a) implies a constant function, "which is in rough agreement with observed facts" (Beckmann, 1973, page 362), but an income elasticity term may be introduced if desired[27].
(3) Density is inversely related to income [equation (11.29b)]. In this model the simplest explanation is that the rich have a strong preference for more space (a high income elasticity of demand for space), perhaps reinforced by conditions (neighbourhood quality, amenities, access to open space, flight from blight, pollution, and high taxes) that make the marginal utility of distance strongly positive. However, even if distance does not enter the utility function (a common assumption in many recent urban models) the inverse income-density relationship still holds. These models assume one transport mode; hence, transport costs from a given location to the CBD are the same for the rich and poor alike. Consequently, the rich must spend more on housing. To satisfy the first-order condition for utility maximization ($p'_r/s^{-1} + t'_r = 0$, where t'_r is the change in transport costs at r), the rich *must consume more space* at a lower rent. Rents must decline with distance from the CBD, otherwise there would be locational disequilibrium since the rich could increase their utility by moving closer in and obtaining the same space at lower cost (since the utility function implies they are indifferent to location).

The relationships between rent, densities, and income are summed up in figure 11.1. y_1, y_2, and y_3 represent a family of hyperbolic isoincome curves. For a given income, high rents are associated with high densities; conversely, low densities are associated with low rents. These relationships may hold cross-sectionally between cities in the same income class, or within a single city. Each isoincome curve could represent a particular income class within a city, thus allowing households with the same income to be distributed over space at different distances from the CBD (an empirical finding ruled out by some of the more restrictive urban theories); or the curves could represent different cities with unequal incomes; or again, the curves could show the path of one city over time. The impact of higher incomes on rents (positive) and on densities (negative) is clear. Another use for the diagram in the multiple-income groups, single-city case is to allow rents to reflect demand for accessibility, and to draw a set of access-space trade-off functions for the individual household. The point of tangency between the lowest trade-off function and the

---

[27] A more general formulation of equation (11.29a) is $p = Ky^{\epsilon_1}(s^{-1})^{\epsilon_2}$. Equation (11.29a) assumes that $\epsilon_1 = 1$ and $\epsilon_2 = 1$. The introduction of income and density elasticities may result in an improved specification of the rent function and may facilitate empirical testing.

appropriate income curve represents the optimal allocation between access (as reflected in willingness to pay the prevailing rent) and space. The slope of the function indicates the strength of the household's preference for accessibility relative to space, that is, a steep function implies a preference for living at the centre of the city.

The standard predictions of continuous models, that the rich live further out, consume more space, spend more on housing, but pay lower rents per unit area, can be obtained by assuming that the slope of the access-space trade-off function is inversely related to the income level. However, the model is general enough to allow the wealthy to have steep functions and to live close in, thus consuming more space than a poorer person at that distance. If the diagram is adaptable to deal with the relationships between rent, density, and income at the level of the individual household, it is possible to generate situations where, subject to the condition that the price at any given location must be the same for all locators, different income groups can live at the same distance at different densities or the same income group can live at different densities and distances. This heterogeneity, not unknown in the real world, is usually ruled out in neoclassical utility-maximization models by restrictive assumptions about identical utilities or about the smoothness of adjustments to competitive equilibrium.

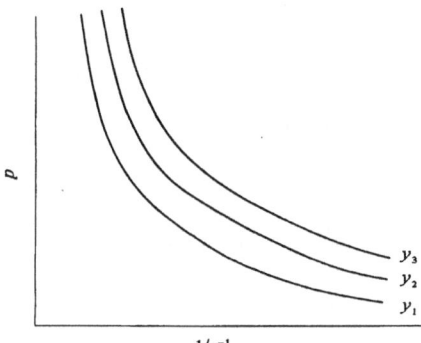

Figure 11.1. Rent, density, and income.

The discontinuous density model

The basic model is similar to the standard urban models in that it generates continuous rent and density functions. To obtain a discontinuous density function in response to continuous changes in incomes and rents, two modifications are introduced. First, the direct rent-density and inverse income-density relationships are qualified. The expression, $1/s^{-1}$, or area per household, measures the gross unit space available. It is assumed that there is a floor to $1/s^{-1}$, defined as a constant $\beta(\beta > 0)$, which sets a maximum degree of overcrowding. It is not total area per person but the excess over this minimum (or, net space is equal to gross space *minus* β) which is influenced by what happens to rents and incomes. Behaviourally, higher rents or lower incomes induce households to tolerate a lower net space than they might otherwise prefer. Thus

$$\frac{1}{s^{-1}} - \beta = \frac{Ky}{p}, \qquad (11.30\text{a})$$

or

$$s^{-1} = \frac{p}{\beta p + Ky}. \qquad (11.30\text{b})$$

At low rents and high incomes, population densities are similar to the basic model case. However, if there is an increase in poverty (falling y), or a substantial increase in rents, densities tend to increase towards a maximum ($1/\beta$). Net space disappears and the city (or the location subject to the income or rent changes) becomes highly congested.

Second, it is assumed that households are willing to spend more on a given amount of housing at low densities. This is an environmental externality (cf Mirrlees, 1972, pages 126-131). The simplest way of taking it into account is by modifying equation (11.29a) so that

$$\frac{p}{s^{-1}} = Ky - \alpha s^{-1}, \qquad (11.31\text{a})$$

where

p/s^{-1} is the housing expenditure (rent per acre divided by households per acre), and

α is a constant (the density disutility coefficient),

or

$$p + \alpha s^{-2} = Kys^{-1}. \qquad (11.31\text{b})$$

These two important refinements can be integrated into the basic model. As a result

$$p + \alpha s^{-2} = \frac{Kys^{-1}}{1 - \beta s^{-1}}, \qquad (11.32\text{a})$$

or

$$(p + \alpha s^{-2})\left(\frac{1}{s^{-1}} - \beta\right) = Ky. \qquad (11.32\text{b})$$

This is, in effect, van der Waals equation, except that $1/s^{-1} = V$, for imperfect physical gases. The original equation was used to show how continuous changes in pressure and temperature may bring about abrupt changes in the volume of a gas (corresponding to liquid, vapour, and liquid-vapour states). Mathematically, equation (11.32b) is a cubic equation:

$$\alpha\beta s^{-3} - \alpha s^{-2} + (Ky + \beta p)s^{-1} - p = 0. \qquad (11.32\text{c})$$

There is a three-dimensional surface, $U = f(s^{-1}, y, p)$, which satisfies this equation. It contains a critical region in its (y, p) plane where there are three values of s^{-1} for each pair of (y, p) values. At the boundary of this region, there are two s^{-1} values for each (y, p) pair. There are no major changes in s^{-1} values as a (y, p) path enters the critical region, but when such a path leaves the critical region there is a discontinuous change in the value of s^{-1}. The (y, p) values at the boundary exit may be called the critical values. The critical values for y, p and the corresponding value for s^{-1}, that is, where the discontinuity occurs, are given in table 11.1, on the assumption of different values for the constants α, β, and K. As pointed out in the note to the table, the critical parameter values can be expressed in terms of the constants.

Table 11.1. Critical values for discontinuities in densities.

Constants			Critical parameter values [a]		
α	β	K	p	y	s^{-1}
20	0·03	0·1	823·0	1975·3	11·11
50	0·05	0·25	747·4	1185·2	6·67
100	0·1	0·4	370·4	740·7	3·33
120	0·15	0·3	197·5	790·1	2·22
200	0·1	0·25	740·8	2370·4	3·33
300	0·2	0·5	277·7	888·8	1·67

[a] Critical values are derived from the following relationships:

$$p = \frac{\alpha}{27\beta^2} \; ; \quad s^{-1} = \frac{1}{3\beta} \; ; \quad y = \frac{8\alpha}{27\beta K} \; ; \quad \frac{Kys^{-1}}{p} = \frac{8}{3} \; .$$

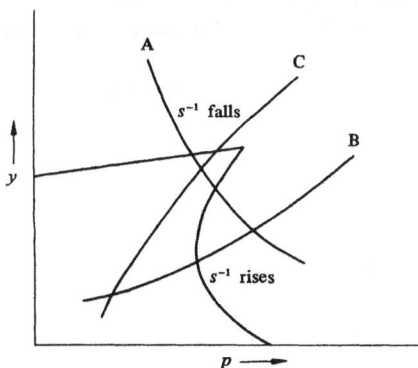

Figure 11.2. Discontinuous density paths.

Figure 11.2 presents three examples of (y, p) paths that cross the critical region from inside. Path A describes a situation where incomes rise and rents fall. This corresponds to two real-world cases: (1) the decline in rents and increase in income with increasing distance from the CBD (the spatial case); and (2) rising incomes and falling relative rents (associated with transportation improvements) over time bringing about markedly declining densities, and discontinuous spatial expansions to cities (the dynamic case). A discontinuous drop in density occurs as the A path moves out of the critical region. In the dynamic case (2) locational inertia prevents sudden jumps in density in real time, but the abstract model remains an approximate representation of reality. Path B describes a situation where income and rents both increase, corresponding to conditions of rapid growth in a city over time. The result is a discontinuous upward shift in densities, as growing demand and high rents lead to congestion and overcrowding. Along path C incomes increase much faster than, and hence offset, rising rents; accordingly, density falls. As implied by the location of the critical region in figure 11.2, if cities are so wealthy or their rental levels so high that their (y, p) paths lie outside the critical region, no discontinuity in densities occurs.

Comparison with the standard neoclassical model

The discontinuous model is not so much a contradiction of the standard urban model as an extension of it. The similarities between the basic model discussed above, the neoclassical model, and its empirical equivalent—the density-gradient model—are clear from the functions shown in table 11.2. For example, s_c^{-1} is much the same as p_o/ky. Since housing expenditure (ky) is rent per acre divided by density per acre (p/s^{-1}), CBD density is equal to CBD rent divided by housing expenditure [from equation (11.29b)]. Also $\exp(-br)$ is comparable to $w_r^{1/k-1}$, since applying a fixed exponent to variable w_r ($w_r' < 0$) is equivalent to applying

Table 11.2. Density functions.

Density gradient:	$s_r^{-1} = s_c^{-1} \exp(-br)$
Neoclassical model:	$s_r^{-1} = \dfrac{p_o}{ky} w_r^{1/k-1}$
Basic model:	$s_r^{-1} = \dfrac{p_r}{ky_r}$

Key:
s_c^{-1} is the density at the city centre
r is the distance from the city centre
b is the slope of density gradient
p_o is the CBD rent
w_r is the fraction of income left after transport costs from distance r
k is the fraction of income net of transport costs spent on rent

a variable negative exponent ($-br$) to the constant, e. The neoclassical and basic models are also similar, since $p_r = p_0 w_r^{1/k}$ and $y_r = w_r y$ (in the case of identical incomes). Since the discontinuous density model is a refinement of the basic model to allow for behavioural responses to congestion, the pressure of high rents and poverty (or affluence), its modifications could equally be introduced into a neoclassical model, though the analytical solutions would be much more difficult.

The model developed here has two major advantages over the standard model. First, the intertemporal version can be used for dynamic analysis, especially to explore the effects of growth in incomes, whereas the neoclassical urban models have all been static. Second, discontinuities in spatial structure are more realistic than the smooth functions of the standard model. The only way to obtain discontinuities in the latter is to introduce housing as a durable good and to allow for discrete sequences in spatial development (see Anas, 1976b).

On the other hand, the model has certain limitations as compared with the neoclassical model. Transport costs and changes in the transportation system are treated only implicitly, whereas the standard model has been rich in its analysis of the land-use requirements of transportation and of road congestion. Although changes in rents and incomes interact with each other to alter the spatial structure, an inevitable consequence of treating them as exogenous is that the interrelationships between income and rents are not as developed as in the standard model. Last, a result of not determining rents and incomes within the model is that it is only a partial model. However, this drawback can be turned to advantage because the theory may be integrated into a much more comprehensive framework for analyzing urban structure and growth as one of several submodels. Also, the simplicity of the approach makes empirical testing much more feasible than with the neoclassical utility-maximization models. How to assess the balance of advantages and disadvantages is unclear. Probably the answer relies on judgment and on the purposes of the model. The claim here is not that the discontinuous model is superior to the standard model, but rather that it offers a different perspective that may have much greater potential than is realized at present.

Policy implications
Although it would not be appropriate to draw direct policy inferences from a simple theoretical model of this kind, the approach is nevertheless suggestive for urban policymakers and planners. The model offers an alternative view of the world to the standard neoclassical model. The latter implies that competitive processes in the land market leads spontaneously, via the smooth operation of market forces, to an 'ideal' spatial structure characterized by continuous gradients and contours. The physical analogies of the discontinuous model, incomes as 'heat' and rents as 'pressure', suggest an element of stress and instability. The refinements introduced in

equations (11.30) and (11.31) express the importance of behavioural influences on residential-location decisions in which spatial choices are constrained or expanded by income and price effects in ways more subtle than those found in the neoclassical model. Although the discontinuities that may result are not necessarily bad—indeed, they may be quite consistent with the heterogeneity of household preferences and the behavioural properties of the housing market—understanding how the spatial structure is moulded by behavioural responses may better equip the planner to deal with pathological situations than a blind faith in the outcomes of deterministic markets.

Even more important is the fact that the model emphasizes the interaction between spatial structure (in particular, density measures) and economic variables. The direct implication is that policies to affect the general economic welfare of the city will have major repercussions on its spatial structure. The functional relationship of equation (11.29a) may be treated as the basis of a policy model in which density is the state-of-system variable, and income and rent are control variables. This provides a contrast to the sole emphasis on zoning, lot-size restrictions, and other physical planning instruments that may not only have distorting effects on the spatial structure but could also have undesirable feedbacks on the welfare and economic performance of the city. The discontinuous model implies that density variations are, to a large extent, the *result* of economic changes, specifically in income levels and in the rental pattern (which is strongly influenced by conditions in the housing market).

It would be going too far to invert the argument by contending that physical planning instruments have no useful role in changing the spatial structure favourably, but this model at least implies that urban policy-makers and planners must also operate on general economic variables—some of which are nonspatial—rather than in the belief that the problems of urban spatial structure can be handled via physical planning controls alone. Actions to raise the level of urban income (and in some cases perhaps to control its growth rate) and intervention in the housing market (to affect house prices and rents) have a claim to be considered along with physical planning measures as instruments of urban policy. Similarly, measures that influence net disposable income and rent levels indirectly, of which the most obvious are transportation innovations and improvements, will also affect the spatial structure via income and price effects. Measures to promote efficiency in housing supply fall into a similar category. There is a difference in emphasis between these inferences and the policy implications of the standard neoclassical model. The latter either equates competitive with optimal equilibrium or, where intervention is justified, recommends pricing strategies (for example, road-congestion tolls) with the aim of changing the slope of the density gradient. Since the scope for pricing strategies is restricted by the limits of political acceptability, there is something to be said for the more varied policy-mix suggested by this alternative approach.

Conclusions
The model developed here offers an alternative to the more familiar, continuous urban models. Although originating in a physical analogy, the model makes economic sense. It may not have general validity, since its special property—the generation of discontinuities in density functions—depends upon income and rent levels having specific values. Nevertheless, only one of a very large number of variations of the basic model has been examined here. Other behavioural functions may be introduced to derive results consistent with economic analogues of some of the other gas equations.

As argued above, a particular virtue of the model is that it is adaptable to either spatial or intertemporal analysis. However, the spatial version raises a problem. One might expect a discontinuity in the density function to break down via the movement of households over the border from the high- to the low-density zone. To a limited extent such an invasion may occur, but the persistence of the discontinuities is indicative of restrictions on mobility. Income constraints may be a problem because lower densities probably involve much higher housing expenditures. In a more complex model there might be feedbacks from densities to rents, breaking down the smoothness of the rent function. This opens up the analytical door to restrictions on competition, such as barriers to 'income mixing' and the existence of 'dual' housing (and spatial labour) markets, which inhibit the proper functioning of rent markets in cities. Residential segregation theory (by income, class, or race) provides additional support for the more abstract model outlined here.

Alternatives to NUE

The criticisms of NUE discussed earlier in this book would be weakened considerably if there were no viable alternatives to NUE available. The purpose of this and the next chapter is to evaluate the major alternatives that have appeared in the literature. In this chapter attention is focussed on noncontinuous models by economists (these were mentioned in chapter 9) and on operational planning models. Whether these are preferred to NUE depends on the assessor's trade-off between operationality on the one hand and theoretical elegance on the other. In the next chapter philosophical outlooks different from those of the neoclassicists are discussed, with most attention given to the critiques of the Marxists and the radical political economists. The theoretical implications of their alternative perspectives are only now beginning to be analyzed by urban economists and other social scientists (for example Farhi, 1973; Scott, 1975a; 1975b; 1976a; Harvey, 1973; 1974; 1975). Although these alternatives to NUE are discussed critically, no attempt is made to pass final judgment on the question of whether they represent more fruitful approaches. It is much too soon to attempt such a judgment. In any event, the choice of methodology is a matter of opinion and taste. However, it is clear that these alternative models deserve serious consideration.

Linear-programming models

In choosing between the theoretically elegant but on the whole nonoperational continuous models and the clumsier, but practical (though still with considerable difficulty), discontinuous models, linear-programming approaches make a strong claim for the latter. One of the earliest and most interesting attempts in the field of residential spatial structure was by Herbert and Stevens (1960). This was part of the famous, or perhaps notorious, Penn-Jersey Transportation Study, and was never applied owing to its heavy data requirements. Nevertheless it is very close to a discontinuous analogue of the standard model (especially the bid-rent version of Alonso, 1964), and is potentially operational.

The model shows how households may be assigned to locations when they maximize their rent-paying ability. Housing needs are satisfied in the market, and location decisions are based on a comparison of the costs of a 'residential bundle' with the budget available. A residential bundle is "a unique combination of a house, an amenity level, a trip set and a site size" (Herbert and Stevens, 1960, page 24), that is, of house type, environmental quality, accessibility, and space consumption. Rent-paying ability is defined as the difference between the costs of the bundle, *excluding site cost*, and the budget allocated by the household to purchase the bundle (this is not dissimilar to the bid-rent concept). The model optimizes locations for households in different groups (for example, income strata) subject to capacity constraints in each residential zone and a given number of households in each group (that is, ensuring that every household is assigned to a site). If rent-paying abilities are maximized, the dual shows that total rents are minimized. Provided that taxes and subsidies are permitted to apply to particular household groups, the linear-programming model replicates a market-clearing mechanism.

The model assumes a finite set of household groups ($i = 1$), of residential bundles ($j = 1$) and of zones ($k = 1$). The other variables are:

b_{ij} the budget allocated by household of group i to purchase residential bundle j;
c_{ijk} the cost to household of group i of residential bundle j in zone k, exclusive of site cost;
s_{ij} the area of site occupied by a household of group i if it uses residential bundle j;
$\hat{s}_k$ the area of land available for residential use in zone k in a particular iteration of the model;
N_i the number of households of group i to be located;
n_{ijk} the number of households of group i using residential bundle j assigned to zone k;
p_k the unit rent in zone k;
v_i the subsidy (or tax) per household in group i.

The primal: maximize rent-paying ability

$$\text{maximize } Z = \sum_i \sum_j \sum_k n_{ijk}(b_{ij} - c_{ijk}) , \tag{12.1}$$

subject to

$$\sum_i \sum_j s_{ij} n_{ijk} \leqslant \hat{s}_k , \tag{12.2}$$

$$n_{ijk} \geqslant 0 , \tag{12.3}$$

and

$$\sum_j \sum_k n_{ijk} = N_i . \tag{12.4}$$

The dual: minimize land rents

$$\text{minimize } Z' = \sum_k p_k \hat{s}_k + \sum_i v_i N_i , \tag{12.5}$$

subject to

$$s_{ij} p_k - v_i \geqslant b_{ij} - c_{ijk} , \tag{12.6}$$

$$p_k \geqslant 0 , \tag{12.7}$$

$$v_i \gtreqless 0 . \tag{12.8}$$

The objective function of the primal is to maximize aggregate rent-paying ability (in the sense defined above). If the model is to be used for dynamic analysis it must be solved iteratively. Constraints (12.2) prevent the consumption of land in each zone from exceeding the land available. Constraints (12.3) are the standard nonnegativity constraints. Finally, constraints (12.4) ensure that all households in each group are located somewhere.

As in many linear-programming models, the dual is easier to handle. The first term, Z', in the objective function is total land rent. At first sight it may appear paradoxical that the dual minimizes land rent whereas the primal maximizes aggregate rent-paying ability. But minimizing site rent implies minimization of returns to landlords; thus this is a competitive solution. Constraints (12.6) ensure that the unit rent on any site does not fall below the rent-paying ability of any household that might locate on that site. As a result, the individual landowner receives at least as much as the highest bidder is willing to pay. However, this creates a difficulty if the group that can bid the highest does not acquire the land because it has an even higher unit rent-paying ability elsewhere. This explains the need for the second term in the objective function—the subsidy variables.

The requirement that all households have to be located makes the subsidy variables necessary; they are required mathematically. For each constraint in the primal there must be a variable in the dual (and for each variable in the primal there must be a constraint in the dual). The land constraints (12.2) are associated with the variables p_k. Similarly the allocation constraints of the household group, constraints (12.4), are associated with the subsidy variables v_i.

However, more important justifications for the inclusion of v_i are economic. Consider the case where all household groups except two have been located and all zones except one have been filled. One of the household groups can outbid the other, and this will set the rent level, but the remaining group has to be located somewhere. Since all occupants of a zone have to pay the same rent, the only solution is for the households that make low bids to be subsidized. This seems a reasonable solution, but it raises yet another difficulty. Since subsidies are given to all members of the group, in the case where households of the same group occupy more than one zone, a subsidy under the conditions described above would also have to be given to households of the same group that have already been allocated, unnecessarily raising their rent-paying ability. There are two easy conceptual solutions to this problem, but they severely restrict operationality. First, household groups could be disaggregated— in the extreme, constraint (12.6) could be placed on each individual household. Second, and more feasible, zones could be disaggregated after a preliminary run of the model in order to subdivide areas shared by two or more household types. Rents could be adjusted by taxing away from landowners the higher rents resulting from the subsidies. Rents would then be equal to rent-paying abilities for all household groups. The greater the degree of zonal disaggregation the closer is the approximation to a continuous model.

Another problem may arise if the household group with the highest rent-paying ability in a zone cannot fill it in a situation where all other households have higher rent-paying abilities in other zones and can be accommodated there. The land constraint (12.2) for the partly empty zone would remain an inequality in the optimal solution; hence the rent should be zero. But the dual constraints require that the land rent should be no less than the rent-paying ability of any household that might locate there. If rent-paying ability is positive, there is a contradiction. However, the conceptual solution is to use a negative v_i—a tax rather than a subsidy— so that the two conditions, zero rent in unfilled zones and unit rent greater than or equal to unit rent-paying ability in all zones, are satisfied.

There are several limitations of this model. They include: heavy data requirements; difficulties of using the model for forecasting purposes because of the problems of obtaining predictions of locating households and available land; restriction of household choice to a single indifference set; optimization over time by iteration may yield different results from a continuous dynamic model; and neglect of locational interdependence. On the other hand the model can be adapted for policy questions such as zoning, renewal, public housing, and segregation (if they have implications for rent-paying abilities).

Several issues merit a brief consideration. One difficulty is that the preference function for housing is more complex than in NUE models owing to the addition of house type and amenity level to the more familiar accessibility and space variables. Since these new characteristics possibly vary more among households than access and space, in addition to the fact that there are now four rather than two (multidimensional) variables, the number of household groups must become large. There is no easy way of identifying preferences, and the most obvious solution— household surveys—is costly, and the results are difficult to interpret. If the model is to be used for predictive purposes, *future* preferences need to be known at future income levels. The assumption of unchanging tastes is too dangerous. Of course, if these difficulties can be overcome, the model permits a much richer treatment of housing preferences than the standard access–space models.

This model also suffers from the familiar drawback of linear-programming models that the optimal solution may be unrealistic in the sense of generating a residential distribution very different from the actual distribution. In the model there will only be as many values of n_{ijk} as there are constraints. If there are K zones and I household groups, the primal problem contains $K+I$ constraints. Therefore there will be no more than $K+I$ nonzero variables in the optimal solution. If at least one household locates in every zone, no more than I zones can have a second type of household. Wilson (1974, page 203) argues that this problem could be met by converting the Herbert-Stevens model to an entropy-maximizing form, a change which could lead to more zonal mixing than the original model.

The way of doing this is simple. Equation (12.1) is now converted to a constraint, with Z no longer being a maximand. Instead, entropy S is maximized by using the standard formulation (see page 200):

$$\text{maximize } S = -\sum_i \sum_j \sum_k \log n_{ijk} \,, \tag{12.9}$$

subject to

$$\sum_i \sum_j \sum_k n_{ijk}(b_{ij} - c_{ijk}) = Z \,, \tag{12.10}$$

and to constraint (12.2) and equation (12.4), where Z is now some number. In addition

$$n_{ijk} = A_i N_i \exp(-\lambda_k s_{ij}) \exp \mu(b_{ij} - c_{ijk}) \,, \tag{12.11}$$

where

$$A_i = \frac{1}{\sum_j \sum_k \exp(-\lambda_k s_{ij}) \exp \mu(b_{ij} - c_{ijk})} \,, \tag{12.12}$$

to ensure that equation (12.11) is satisfied, and λ_k and μ are the Lagrangian multipliers associated with constraint (12.1) and equation (12.10) respectively. A_i, λ_k, and μ would be solved iteratively. The value of Z would be chosen so that the values of n_{ijk} reflect the real-world distribution. Since the $Z^{\max}$ of the original model could also be calculated, the expression $(Z^{\max} - Z)$ would measure the degree of suboptimality of the actual allocation. Similar conversions would be necessary to alter the dual in order to identify actual housing expenditures rather than minimization of rents (see Senior and Wilson, 1974, pages 165-166).

To make the model more realistic in other ways—such as taking account of the pull of major employment centres on residential location, traffic congestion costs, crowding impacts on new location decisions, and the influence of changes in land use on amenity level—is not difficult conceptually, but the price to be paid in operationality is that the model then becomes a nonlinear-programming model.

In comparing the advantages of the linear-programming and the continuous models, Herbert and Stevens argue that "the linear programming model may actually be the more realistic of the two. A truly continuous model assumes a level of information and sophistication on the part of both landowners and households which is not likely to exist in practice. From the landowners' point of view it is hard to imagine a real situation where there are marked variations in residential rent among contiguous sites; from the households' point of view, it seems unlikely that contiguous sites will be regarded as having distinctly different locational cost and advantages" (Herbert and Stevens, 1960, pages 31-32). Moreover the discontinuous rental surface that results from the discrete zones of the model may be more realistic than the smooth, differentiable surface of NUE models. For instance, it would be surprising to find smooth rent gradients in a multicentric metropolitan area with uneven topography, mixed land uses, and a nonubiquitous transportation system.

Wheaton (1974a) has recently argued that discontinuities in the form of unequal rents in adjacent zones are not so much the product of the discreteness of the model as an indication that the solution is not a true equilibrium. The values of v_i in the model are, on this view, merely a device to obtain a solution, and the correct solution for locational equilibrium is obtained when all the values of v_i are zero, that is, when bid rents equal actual rents. The reason why this solution is not generated by the model is because of the confessed limitation of single, indifference sets. Rent-paying abilities cannot be determined exogenously, as in the Herbert-Stevens model, but have to be created in the market by arriving, via convergence, at an endogenously determined set of utility levels, so that bid rents yield a solution to the linear programme that has zero subsidy values.

An alternative, more general, linear-programming urban model has been developed by Mills (1972b), in which an efficient spatial allocation within the city is obtained by minimizing the costs of required production targets and transportation. Although cumbersome compared to the elegance of the continuous models, the method has some interesting features. It treats the city in a discrete fashion by dividing the urban area into square grids. It takes account of the fact that the city can grow upwards as well as outwards, and allows capital to be substituted for land by the construction of tall buildings. The model includes several production goods that are exported. Unit production costs may change with building height, with the direction of change depending on input-output coefficients, and on relative factor prices; this allows economies (and diseconomies) of scale to be explored within a linear-programming framework. Assumptions are made to ensure that commuters do not commute away from the city centre, and this implies that housing is more decentralized than production. The transportation system is allowed to reflect congestion by allowing the marginal-transportation-cost function to rise in a stepwise fashion. The transportation system is considered a potentially weak link from the point of view of achieving an efficient spatial allocation, since it has to be provided by the public sector, which is unlikely to optimize its price structure, especially where most travel is by automobiles.

Preliminary tests of the model had mixed results. Slight variations in assumptions (about input-output coefficients and transport costs) could bring a city from one extreme (very little commuting, with most employment in the housing zones) to another (heavy commuting, segregation of goods production and residences, but low shipments of goods), though this probably reflected the restrictive assumption of one export good in the empirical tests. On the other hand the optimal level of transport congestion was insensitive to parameter changes, with transport costs in the city centre five times as high as on the periphery. Buildings were high close-in, but fell off rapidly. The shadow price of urban land also fell off rapidly, though it was fifty to one hundred times greater in the city centre than on the periphery.

Extensions of the model might include: fixing some land uses (for example, historically determined uses) exogenously; modifications in the production assumptions (for example, allowing for total consumption and intermediate goods production); permitting goods to be shipped from suburban locations rather than via the city centre (see Hartwick and Hartwick, 1974, discussed below); and taking account of institutional constraints such as zoning, building-height conditions and property taxes. Some of these refinements might complicate the model beyond the feasibility of empirical testing. Even the model outlined would involve about one hundred inequalities and almost two thousand variables to represent a city of one million people. Nevertheless, empirical analysis is one of the primary objectives of this approach and it is much more likely an outcome than if neoclassical utility maximization models are used.

In a recent paper Hartwick and Hartwick (1974) have extended Mills's (1972b) linear-programming model to deal with multiple centres and intermediate goods. They also assume a geographical area divided into square grids, and it is treated as an agricultural plain with agricultural land rent given. The city develops when the area is required to produce a bill of goods for export to the rest of the system. Production uses three inputs (land, labour, and capital), apart from the transportation sector which is assumed to require land and capital only in order to avoid the problem of having to determine commuting patterns for workers in that sector. A range of techniques is available, each associated with a different capital–land ratio. This allows for multistorey buildings. The multiple centres of the city are assumed to be transportation nodes from which all the city's exports flow.

The primal objective function is to assign all nonagricultural activities (including housing and transportation) to each zone so as to minimize total resource and transportation costs, subject to meeting a predetermined bill of final demands and to the land demands of all activities in each zone not being greater than the land available. Some zones may be empty (that is, left to agriculture) in the optimal solution. The constraints require: that each zone be in commodity-flow equilibrium; that the export of each good must at least satisfy final demand (this may be zero, which makes the good solely an intermediate good); and that the supply of transportation must be equal to the demands for it.

The dual of the linear programme is to satisfy the constrained maximum of the value of urban output net of land values. The major constraint is that profits in all activities must be nonpositive. This makes it unprofitable for any activity to change location, and hence is required for locational equilibrium. Household equilibrium is obtained in the same way as in continuous models. Each worker is associated with one housing unit and has the same real income: this means that additional transport costs will be exactly offset by lower transport costs. These assumptions eliminate the allocation of households in space from the model, or rather, the housing and labour sectors are treated as a unit.

The Hartwicks develop a numerical simulation of the model with the use of four commodities (including housing), plus transportation and agriculture, five techniques (allowing up to five storeys) and two transportation nodes, allocated on a 9×11 rectangular grid. The results were reasonably realistic, generating the expected spatial differentials in capital-land ratios and land rents, with gradients around the two nodes. One good, which was assumed to require much lower transportation services, was much more suburbanized than the rest. The admitted drawback of the model is its inability to explain multiple centres in terms of agglomeration economies. There are ways for dealing with this, even within a programming framework, but none is satisfactory.

Despite their lack of beauty, the linear-programming models offer compelling advantages: the ability to deal with two-dimensional space in a nontrivial way; production is endogenous rather than exogenous (these two advantages explain the limited success with modelling major subcentres); attention to employment as well as to residential densities; and the design of a general framework within which institutional constraints can be accommodated. These gains are achieved without having to sacrifice important model requirements such as taking account of the congestion effects of transport and the derivation of a rent function. Of course, there are offsetting disadvantages: sacrifice of theoretical purity; the need for data in order to derive any conclusions; and the fact that linear approximations to changes in cost functions remain approximations. Nevertheless this line of inquiry deserves more attention than it has received hitherto.

Simplicial search algorithms

MacKinnon (1974) has applied the idea of using fixed-point algorithms for solving nonlinear general equilibrium systems to urban analysis. The method is via simplicial search algorithms. A simplicial subdivision is obtained by dividing up an n-simplex into subsimplices with the same properties as a simplex (n vertices, n faces, and $n-1$ dimensions). A *regular* subdivision can be fully described by a single number (if the coordinates of every vertex in the subdivision are multiplied by a number D, the new coordinates are all integers, and for each vertex they sum to D). A labelling rule allows an integer label (from 1 to n) to be assigned to each vertex in the subdivision, provided that no zero coordinate is given a number. Since each vertex of the original simplex has only one nonzero coordinate, the vertices of the original simplex must have the labels 1 to n. From Sperner's Lemma, every properly labelled simplicial subdivision must contain an odd number of completely labelled subsimplices (subsimplices whose vertices all have n labels). Paths through a subsimplex must terminate either in ends or another start, and there is an odd number of starts (and ends).

Why are simplicial search algorithms useful? Consider a function F which maps from a vector x into another vector x^*. The elements of x are nonnegative and sum to unity, as do the elements of x^*; thus x and x^* both lie on a simplex. By the Fixed Point Theorem there must be at least one vector $\bar{x}$ such that $F(\bar{x}) = \bar{x}$. This can be found easily if the function is differentiable and concave or convex. If not, a simplicial search algorithm may be used. This involves choosing some subdivision of the simplex and labelling the vertices according to the rule:

$$L(x) = K, \quad \text{if} \quad F_K(x) \leq x_K, \quad \text{and} \quad x_K > 0. \quad (12.13)$$

A vertex x is given the label K if the Kth component of $F(x)$ is not greater than the Kth component of x. If a vertex qualifies for more than one label by this rule, one may be chosen by any convenient method. There is an odd number of completely labelled subsimplices. If F is continuous, or D approaches infinity, then for every vertex of each subsimplex, $F_K(v) \leq v$ for every K, because the distance between them goes to zero. Hence $F_K(x') \leq x'_K$ for all K. Thus $F(x') = x'$, and x' is a fixed point of F.

This is helpful in the solution of general equilibrium-price systems. Such a system can be reduced to a number of excess-demand functions, $\phi_i(P)$, where P is a vector of nonnegative prices. At equilibrium,

$$\phi_i(P) \leq 0, \quad \text{and} \quad \phi_i(P)P_i = 0, \quad \text{for all } i. \quad (12.14)$$

The excess-demand functions are usually homogeneous of degree zero in prices. Thus it is possible to normalize prices so that $\sum_{i=1}^{n} P_i = 1$. Hence the price vector is an n-simplex. A labelling rule can be adopted:

$$L(P) = K, \quad \text{if} \quad \phi_K(P) \leq 0 \quad \text{and} \quad P_K > 0. \quad (12.15)$$

If a vertex can have more than one label, a subsidiary rule is chosen. Any rule leads to equilibrium, with a large D, but the choice of rule determines how long it takes. A procedure, called the 'sandwich method', can be used as one particularly useful algorithm: artificial labels can be used, and it may be more efficient to conduct the search on an $(n+1)$-simplex. This can simplify computation since it allows the same problem to be attempted many times (for different values of D) at small additional cost.

Applying this to a theoretical city, MacKinnon divides the city into a number, R, of concentric rings. Assuming one nonland good, the model contains $R+1$ prices which can be constrained to lie on a simplex. The solution is to find a vector of prices that achieves equilibrium (no individual can make himself better off by moving or by changing his consumption), and it is possible to achieve this in the neighbourhood of a completely labelled subsimplex via a choice of labelling rule. As in the traditional models, therefore, the individual utility function and its constraints are the most important element. The labelling rule involves assigning households to their most preferred location until the land runs out, then to their second preferred location, and so on. A draback is that the model generates an approximation to equilibrium (depending on the value of D) but not an optimum.

An interesting result is that in the case of two income groups with different incomes, different utility functions, and possibly even different transport modes, a kink develops in the rent-distance function—a finding not observed in most continuous models. In some results MacKinnon even found one group split in terms of location—one part located near the city centre and the other part located on the fringe. He suggested that a possible assumption to obtain this result might be if the income elasticity of demand for housing was less than unity (beyond a threshold income) and if leisure was an increasing function of income. A further finding was the interdependence between the welfare of the two groups. For example, raising the incomes of the rich had a negative impact on the poor by squeezing them into a smaller area (even though the area of the city expands).

The advantage of this approach is that limitations on particular locational assignments can be introduced (zoning lot restrictions, racial rent differentials, and so on). Also, more complex transportation systems can be handled than in the continuous models. On the other hand some of the restrictions of the latter apply, such as neglect of the locational interdependence between producers and consumers or the computational costs of simulating a two-dimensional city. Nevertheless this class of models makes an interesting alternative.

The possibility of cumulative disequilibrium

Among the alternatives to NUE, one group of models stands out as offering a striking contrast in terms of predictions, but with an even simpler conceptual framework. These are the so-called cumulative disequilibrium models (Oates *et al.*, 1971). They are based on the idea that many dynamic forces affecting urban change are of a cumulative nature, thereby stressing positive feedback loops rather than the negative feedbacks characteristic of equilibrium economics. Although these models are frequently used to illustrate cumulative deterioration—such as the progressive decline of central cities as a result of the flight to the suburbs, they are equally relevant to the exploration of cumulative growth—such as how agglomeration economies may lead to faster growth in the larger cities in the urban system. On the other hand the model is much more general than this implies, since the result of cumulative disequilibrium depends on parameter values. With different parameter estimates the system will converge towards an equilibrium. Another advantage of the approach is that the basic idea can be adapted to a wide variety of specific models. The examples used below are merely simple, illustrative examples. The model has two other attractive features. It can be applied to problems arising at many levels of areal aggregation (for example, individual neighbourhoods, the central city, and the urban area as a whole). It is easy to introduce policy variables so that the framework might be used as a guideline for corrective action.

The Baumol blight model consists of two linear equations:

$$Y_{t+1} = r - s\delta_t, \qquad (s > 0), \tag{12.16}$$

$$\delta_t = u - vy_t, \qquad (v > 0), \tag{12.17}$$

where
y is income, and
δ is a measure of deterioration.
Thus an increase in deterioration leads to a fall in future income, and falls in income induce further deterioration. By substitution

$$y_{t+1} = r - su + svy_t, \tag{12.18}$$

or

$$y_{t+1} = a + by_t, \tag{12.19}$$

where $a = r - su$, and $b = sv$.
To find the equilibrium value, y_e, of y, equation (12.19) is solved for $y_e = y_{t+1} = y_t$. The result is

$$y_e = \frac{a}{1-b} = \frac{r-su}{1-sv}. \tag{12.20}$$

Before considering the properties of this type of model, two other examples may be briefly considered. The first is a similar model in which urban income is determined by the level of urban services. The circular system might be represented by

$$y_{t+1} = a + b\sigma_t, \tag{12.21}$$

$$\sigma_t = c + d\tau_t + \eta E_t, \tag{12.22}$$

$$\tau_t = f + gy_t, \tag{12.23}$$

$$\eta_t = h + iy_{t-1}, \tag{12.24}$$

Alternatives to NUE

where
σ is the level of urban services,
τ is the tax base, and
η is the education level of citizens.

The argument is that services determine the future urban income level by attracting high-income population, the service level is dependent upon the tax base and the present education level (a surrogate for community involvement), whereas the tax base and education depend on current and past income respectively. As the income level changes, it will have direct repercussions on urban services that, in turn, will affect future income. Substituting equations (12.22), (12.23), and (12.24) into equation (12.21), the result is

$$y_{t+1} = a + bc + bdf + beh + bdgy_t + beiy_{t-1} , \tag{12.25}$$

or

$$y_{t+1} = j + ky_t + zy_{t-1} , \tag{12.26}$$

where
$j = a + bc + bdf + beh$,
$k = bdg$, and
$z = bei$.
In this case

$$y_e = \frac{j}{1-k-z} . \tag{12.27}$$

The second example is a simple model of urban growth. The growth rate, y', depends on agglomeration economies, A, which are a function of city size, P. City size, or the urban population, is determined by the number of jobs, E, which is a function of the past growth rate (on the hypothesis that rapid growth attracts new job-creating nonresidential locators). The system may be described as

$$y'_{t+1} = m + nA_t , \tag{12.28}$$

$$A_t = o + pP_t , \tag{12.29}$$

$$P_t = q + rE_t , \tag{12.30}$$

$$E_t = s + ty'_t . \tag{12.31}$$

By substitution

$$y'_{t+1} = m + no + npq + nprs + nprty'_t , \tag{12.32}$$

or

$$y'_{t+1} = w + xy'_t , \tag{12.33}$$

where
$w = m + no + npq + nprs$, and
$x = nprt$.
Of course,

$$y'_e = \frac{w}{1-x} \,. \tag{12.34}$$

There is nothing sacrosanct about any of these models. The reader may easily devise alternatives more suited to his or her taste should any or all of them prove unsatisfactory. To explore the properties of this type of model, let equations (12.20), (12.27), and (12.34) be treated as specific cases of a more general function

$$y_e = \frac{\alpha}{1-\beta} \,. \tag{12.35}$$

The general solution of the model is

$$y_t = (y_0 - y_e)\beta^t + y_e \,, \tag{12.36}$$

where y_0 is the initial value of y_t.

Cumulative growth or decline depend on the value of the parameter β and on whether y_0 is greater than, equal to, or less than y_e. For $y_e > 0$, then $\beta > 1$, $\alpha < 0$, or $\beta < 1$, $\alpha > 0$. Cumulative growth will occur if $y_0 > y_e$ and $\beta > 1$. There is cumulative decline if $y_0 < y_e$ and $\beta > 1$. Conversely, if $\beta < 1$, the process will be convergent. If $y_0 > y_e$, income will fall towards the equilibrium level, whereas if $y_0 < y_e$ income will rise towards its equilibrium. Whether the system is stable or not is simply a question of whether or not β is less than or greater than unity.

The models discussed here are very simple in the sense that they are linear. It is possible to develop more complex systems involving nonlinear difference equations. In such models changes in the values of the main variables could bring about changes in their equilibrium values. In the linear models, however, the only way to stop a cumulative process when this is predicted is by altering the values of the parameters themselves. For instance, Oates *et al.* (1971) allow the parameters, s and u, in equations (12.16) and (12.17) to be influenced by policy variables—the levels of taxes and urban service expenditures [bringing their model closer to that represented in equations (12.21) to (12.27)]. High taxes accelerate the flight to the suburbs, whereas (for a given income level) the level of blight varies inversely with public spending. This extension permits a comparison of the relative effectiveness as a blight-arresting measure of increasing spending or reducing taxes.

Although the empirical potential of this type of model is considerable, very few empirical analyses have been undertaken. Oates *et al.* (1971) present a few 'naive' results, Bradford and Kelejian (1973) develop a simple econometric model of the flight to the suburbs, which owes much to Baumol's theory but uses traditional multiple-regression analysis rather than difference equations (an enforced choice because of the reliance on 1960 and 1970 Census data), and Kumar-Misir (1974) tests a cumulative causation model of regional growth on Canadian data; however, this is a meagre haul. Why there have been so few follow-ups to the Baumol model is difficult to understand. One reason may be its lack of appeal to a profession dominated by equilibrium models from which, as a result of competitive processes, palatable outcomes are derived.

The major difference between this approach and NUE models is that it emphasizes dynamic analysis rather than the static long-run equilibrium framework of NUE. This makes Baumol's models especially appropriate for the analysis of rapidly growing or stagnating cities. NUE models can cope with these conditions only via very inadequate comparative statics exercises.

The Lowry model

Of the several alternatives to NUE, none has received the attention or the range of applications given to the planning model first developed by Lowry (1964). The approach of the Lowry model provides a striking contrast to that of NUE, and a comparison illustrates the advantages and disadvantages of both. The main virtue of the Lowry model is that it has been applied with a reasonable degree of success and computational efficiency. It is capable of dealing with a high degree of spatial disaggregation—hundreds of zones if the computer capacity can cope with inverting the large matrix involved. On the other hand, it is weak theoretically. The shaky economic-base model provides its theoretical underpinnings. The spatial interdependencies between residences and workplaces, and between both of these and service centres are determined by gravity and accessibility-potential relationships that are crudely empirical. Some critical variables in an urban spatial-structure model, such as the location of basic employment workplaces, are determined exogenously. However, the sacrifices in theoretical purity are compensated by the fact that the model can be applied, and has proved itself to be helpful to urban planners and policymakers.

There is no attempt here to cover comprehensively the mound of literature on the Lowry model. The main references used are Lowry (1964), Garin (1966), Goldner (1971), Batty (1971; 1972), Kendrick (1972), and Wilson (1974). Similarly, the comments offered here refer to the basic model rather than to its many elaborations and extensions. Also, it is not intended to review its many applications (some of these have been discussed in the survey papers by Batty and by Goldner). The priority here is to describe the model briefly and to assess its advantages and drawbacks compared with NUE and other more standard economic models.

The model is very simple in form. A spatial variant of the economic-base multiplier is used to generate the spatial distribution of total employment from the location of basic employment, where the latter is determined exogenously. The residential distribution of population is then generated by the distribution of workplaces, inverting the common concept of the journey-to-work into the less familiar journey-to-home. The spatial allocations are made subject to certain constraints: a land-use constraint; a maximum limit on population density in each zone; and a minimum threshold scale of activity for types of service establishments. In view of the simultaneous determination of population and employment, and of the inequality constraints, the model is solved iteratively. However, it is simpler to express in matrix form and, in some circumstances, solution by matrix inversion rather than by iteration is possible.

The metropolitan economy is assumed to consist of $m+1$ sectors, m nonbasic or service sectors ($k = 1$), and one aggregate basic sector, b, and for the purpose of spatial allocation it is divided into n zones ($i = 1$). Let E be an n-vector of zonal total employment, E^b an n-vector of basic employment, and E^k an n-vector of zonal employment in service sector k. Then

$$E = E^b + \sum_{k=1}^{m} E^k . \tag{12.37}$$

The scale and spatial distribution of basic employment is determined exogenously, so that

$$E^b = \overline{E^b} . \tag{12.38}$$

The distribution of service employment depends upon the relative importance of home-based and work-based shopping (or more generally, service) trips.

$$E^k = \hat{A}^k B^k H + \hat{F}^k G^k E , \tag{12.39}$$

where
H is an n-vector of households,
$\hat{A}^k$ is the diagonal matrix of workers required in service sector k per household by zone,
B^k is the sector k demand spatial-distribution matrix for households,
$\hat{F}^k$ is the diagonal matrix of workers required in sector k per worker by zone,
G^k is the sector k demand spatial-distribution matrix for workers.

The right-hand side of equation (12.39) reflects employment in sector k. The first part reflects home-based whereas the second part measures work-based employment. A clearer understanding of equation (12.39) may be obtained by pointing out that it is a more general matrix version of the following equation in Lowry's original formulation:

$$E_j^k = b^k \left[\sum_{i=1}^{n} \left(\frac{c^k H_i}{T_{ij}^k} \right) + d^k E_j \right] , \tag{12.40}$$

where
T_{ij}^k is a trip-distribution index between i and j for sector k,
c^k and d^k measure the relative importance of homes and workplaces as origins for a particular type of shopping, and
b^k is a scale factor to adjust sector employment in each zone to the metropolitan total.

The form of equation (12.40) implies that work-based shopping trips are pedestrian trips so that the only relevant origins are in zone j. Home-based trips may be longer, but are subject to a 'gravity effect': the likelihood of a shopping trip from i to j diminishes with intervening distance.

The location of households is a function of the location of employment, thus

$$H = \hat{C}DE , \qquad (12.41)$$

where
$\hat{C}$ is a diagonal matrix of households per worker by zone (that is, a form of reciprocal activity rate), and
D is a spatial-distribution matrix for distributing the labour force into households by residential zone (Batty, 1972, calls this a journey-to-home probability-distribution matrix).

The original Lowry version of equation (12.41) is

$$H_j = g \sum_{i=1}^{n} \frac{E_i}{T_{ij}} , \qquad (12.42)$$

which implies that the number of households in each zone is a function of that zone's accessibility to employment opportunities. The coefficient g is a scale factor to satisfy the requirement that the sum of zonal populations must be equal to the total population as determined by total employment.

To solve the model from the point of view of spatial distribution of households and employment, equation (12.37) is rearranged to give

$$E^b = E - \sum_{k=1}^{m} E^k . \qquad (12.37a)$$

Substituting equation (12.41) into equation (12.39) yields sector employment as a function of total employment. If the following definitions are made:

$$\hat{A} = \sum^{m} \hat{A}^k , \quad B = \sum^{m} B^k , \quad \hat{F} = \sum^{m} \hat{F}^k , \quad \text{and} \quad G = \sum^{m} G^k ;$$

it is then possible to aggregate over all service sectors. Substituting the revised versions of equation (12.39) into equation (12.37a) gives

$$\begin{aligned} E^b &= E - (\hat{A}B\hat{C}D + \hat{F}G)E \\ &= [I - (\hat{A}B\hat{C}D + \hat{F}G)]E . \end{aligned} \qquad (12.43)$$

Inverting the matrix allows total employment to be determined as a multiple of basic employment, so that

$$E = [I - (\hat{A}B\hat{C}D + \hat{F}G)]^{-1}E^b . \qquad (12.44)$$

This is, of course, a spatial version of the economic-base multiplier, and bears a far from coincidental similarity to the interregional input-output multiplier (see Richardson, 1972, pages 43-52). The metropolitan population distribution may be obtained by substituting equation (12.44) back into equation (12.41), that is

$$H = \hat{C}D[I - (\hat{A}B\hat{C}D + \hat{F}G)]^{-1}E^b . \qquad (12.45)$$

Since the Lowry model is more than a demographic and economic activity model, but is intended as a land-use planning model as well, land uses have to be considered explicitly. The model identifies four land uses: unusable land, basic-employment use, service-employment use, and residential land (denoted by 1, 2, 3, 4 respectively). Thus

$$A_j = \sum_{z=1}^{4} A_j^z , \qquad (12.46)$$

$$A_j^1 = \bar{A}_j^1 , \qquad (12.47)$$

$$A_j^2 = \bar{A}_j^2 , \qquad (12.48)$$

and

$$A_j^3 = \sum_{k=1}^{m} w^k E_j^k , \qquad (12.49)$$

where
A_j is the land use in zone j, and
w^k is the exogenously determined employment-density coefficient for sector k (that is, land requirements per worker).

Since the amount of unusable land and land needed for basic employment are given exogenously, whereas service-sector land use is a function of service employment, residential land use is determined residually.

The constraints of the model may be expressed as follows:

$$A_j^4 \geqslant 0 , \qquad (12.50)$$

$$H_j \leqslant s_j A_j^4 , \qquad (12.51)$$

$$E_j^k \geqslant X_{E^k} , \quad \text{or} \quad E_j^k = 0 . \qquad (12.52)$$

Constraint (12.50) rules out the possibility of residential land use being negative. Constraint (12.51) prevents zonal densities from exceeding a maximum, where s_j is the maximum number of households permitted per unit of residential space. This may vary from one zone to another, since residential zoning standards do vary among different sections of a city. Finally, to reflect minimum economies of scale, constraint (12.52) prevents the size of any service activity within a zone from falling below a minimum threshold level. Although these constraints are easy to understand they complicate the analytical solution, especially in the last example which takes an 'either-or' form. In general, they tend to tip the balance in favour of iterative methods of solution rather than the more convenient matrix-inversion approach.

This is a brief summary of the Lowry model. What assessment can be made of it? The reliance on 'basic' and 'retail' divisions of economic activity should not be interpreted as a literal acceptance of the *export*-base model. Lowry uses these terms for convenience, and accepts that 'site-oriented' and 'residence-oriented' activities might be more exact terms. Nevertheless, he argues that these " 'export' industries are relatively unconstrained in local site selection by problems of access to local markets, and their employment levels are primarily dependent on events outside the local economy. Consequently, they have been treated as exogenous to the model, as activities whose locations and employment levels must be assumed as 'given' " (Lowry, 1964, pages 2-3). Whereas it may be reasonable on practical grounds to ignore the local market demand of basic activities, it is a weakness to have to neglect economic influences on their site location (for example, the trade-off between central-city agglomeration economies and transport cost and rent savings that differ from one type of basic activity to another). This problem is assumed away in the original Lowry model by treating basic activities as a single industry.

The nonmarket character of the Lowry model is reflected in some of its other features. For instance, the maximum-density constraint says nothing about the influence of higher densities that are lower than the maximum on residential site choice. The gravity relationships implicit in equations (12.40) and (12.42) are poor demand equations. Empirical regularities of the type given in the gravity model may facilitate testing and policy relevance, but they are weak substitutes for prices and associated production, and demand and profit functions that lie at the heart of economic models. However, Lowry himself was careful to point out how his model and more traditional locational microeconomics could be made compatible (Lowry, 1964, pages 20-54), and justified his choice of the gravity model simply because it was "much easier to build and cheaper to operate" (*ibid*, page 23).

The NBER Urban Simulation Model
From the viewpoint of progress in urban economics, it would be desirable to develop analyses that blend the different approaches—reliance on economic theory as in NUE, making the best use of empirical data as in econometrics, and replicating the behaviour of the urban economy as in simulation models. In fact, one study that takes steps in this direction is the National Bureau of Economic Research (NBER) Urban Simulation Model developed by John Kain and his associates (Ingram *et al.*, 1972).

The model builders also sought to justify their model as a tool in policy evaluation, though it would require more work before it was really useful for this purpose.

The NBER model focusses almost exclusively on the housing market. It deals explicitly with market behaviour rather than empirical and statistical regularities that indirectly shed some light on what happens in the market. The NBER model gives detailed attention to demand, supply, and price formation. The types of market behaviour directly simulated include: household-mover decisions; price determination by housing type and location; determination of housing type and location chosen by new and moving households; filtering of the housing stock, renovation and conversion; new housing construction; and changes in interzonal travel-to-work patterns. In order to simulate these processes the model is divided into seven submodels dealing with employment location, movers, vacancies, demand allocation, filtering, supply, and market clearing. The first application of the model was in the Detroit metropolitan area, which was divided into nineteen workplace and forty-four residential zones. The workplace zones were centrally located, in fact an aggregation of the thirty-two inner residential zones; the twelve most peripheral residential zones were assumed to contain no workplaces. The model dealt only with employed households, and assumed one worker per household. Each household was classified as belonging to one of seventy-two household types defined in terms of family size, income, education, and age of the head of the household, and lived in one of twenty-seven housing types defined in terms of structural type, number of rooms, quality, and lot size. Travel to work was by one of two transport modes, depicted by interzonal travel time and cost. The scale and location of metropolitan employment were determined exogenously.

Subsequently, the data base was transferred to Pittsburgh to enable more use to be made of neighbourhood quality characteristics in housing submarkets. Pittsburgh I contained twenty workplace zones and fifty residential zones (forty-one high-quality and nine low-quality). Each of twenty different housing types could be provided in either a high- or low-quality zone, so that there are forty housing types altogether. Also, the Pittsburgh data base provided more detail on the prices and other characteristics of individual dwelling units, whereas most of the Detroit data had to be inferred from statistics given in the Census tract. A more developed version of the Pittsburgh model (Pittsburgh II) would go further in correcting some of the deficiencies of the original model. First, there is a need for more variables relating to neighbourhood quality—socioeconomic status and residential density as well as the quality of average dwellings. This modification has the operational advantage of reducing the maximum size of the linear-programming problems to be solved by the market-clearing submodel. Stratifying residential zones by neighbourhood quality allows the number of residential zones to be increased with but a minor increase in model running time. Second, it would be possible to remedy the omission of ignoring racial differences (nonwhite households were excluded in the Detroit version of the model) in one of several ways: by making race a dimension of neighbourhood differences; or by exogenous specification of black and white neighbourhoods (a reasonable simplification in highly segregated cities); or by price markups and discounts if households of a particular race wished to live in a particular neighbourhood. Another improvement would be to take account of schools and the provision of other public services as an influence on the housing market, but empirical research in this area has been very limited (though a spate of studies was triggered by research by Oates, 1969). It would be possible to make nonbasic employment endogenous, perhaps as a function of basic employment and population as in the Lowry model. Another refinement would be to include a submodel for the land market, with the possibility of dealing with residential and nonresidential competition for sites. Once some of these extensions were included it might be feasible to use Pittsburgh II for policy analysis.

Real data from actual cities were used to obtain econometric estimates of some of the major parameters. However, the cities were test beds for evaluation of the structure of the model, rather than the model being used to represent Detroit or Pittsburgh. Indeed, Pittsburgh is so atypical (owing to its topographical irregularities, relative economic stagnation, dependence on heavy industries, and dispersed employment distribution) that it is probably a poor test bed, but its choice was justified by the availability of rich data.

The NBER model attempts to correct two major deficiencies of standard urban economic models, including the NUE models. These are the assumption that production is undertaken at a single location (monocentricity) and the preoccupation with long-run equilibrium solutions, ignoring capital stocks and the consequences of heterogeneous, durable and immobile capital. Multiple employment zones are introduced because a relatively low proportion of economic activity (in the case of manufacturing typically less than fifteen percent) is located in the city centre. Since residential capital is not malleable, cities never reach a long-run equilibrium. In standard theories the assumption of long-run equilibrium, in effect, assumes that "either cities are destroyed every night and rebuilt the next morning or that households live in house trailers that are relocated daily" (Ingram *et al.*, 1972, page 16). Thus the NBER model does not attempt to reproduce a long-run equilibrium. Instead, the model yields estimates of the desired demand for housing by type and location for each time period. The existing stock is modified by maintenance, renovation, repair, and new building. Also, the prices determining desired demand in each time period are not the long-run equilibrium supply prices but a set of expected market prices that reflect the existing housing stock.

The heterogeneity of the stock is fully taken into account by dividing the housing market into a large number of submarkets, each matching a distinct housing type. The more complex problems of interdependencies in urban housing markets (for example, externalities in housing consumption and production, racial segregation, the supply of public goods) were not adequately dealt with in the Detroit Prototype of the NBER model, but one of the aims behind Pittsburgh II was to improve the representation of these effects.

The NBER model is grounded in standard microeconomic theory. Households choose the type and location of housing that maximizes their utility, housing suppliers build and convert to maximize profits, and maintenance decisions are consistent with investment theory. On the household side the model uses demand functions for different dwelling types rather than tracing these back to the original household utility functions which are so difficult to specify. However, in the case of dwelling units, profit maximization can be dealt with directly. Firms are assumed to build at the most profitable location by using the most efficient (least cost) technology. This is obtained in the model by comparing expected prices of houses by type and location with exogenously determined construction costs. Bringing these two sides together is handicapped by the absence of a market-clearing algorithm to generate market prices for use as determinants both in the demand and in the supply relationships. The solution adopted in the NBER model was to use a linear-programming algorithm that replicates the market process with market prices obtained from the shadow prices (dual variables). Thus the NBER model has three major components—a demand sector, a supply sector, and a price-formation sector—combined recursively to replicate the operations of the housing market during each time period. The model does not generate a long-run equilibrium solution but is, in effect, an adjustment model applied period by period. Decisionmakers attempt to move closer to equilibrium, but within each period there are forces in operation disturbing equilibrium, so that "the housing market is perpetually chasing a moving target" (*ibid*, page 30).

The seven submodels mentioned above fit into the above framework in the manner summarized in figure 12.1. The employment-location and the movers- and demand-allocation submodels form the demand sector, whereas the vacancy, filtering, and supply submodels make up the supply sector. Both sides are then brought together in the seventh, critical submodel—the market-clearing submodel. However, another way of splitting up the submodels is to group them according to whether they represent demographic behaviour or housing-market operations (see figure 12.1).

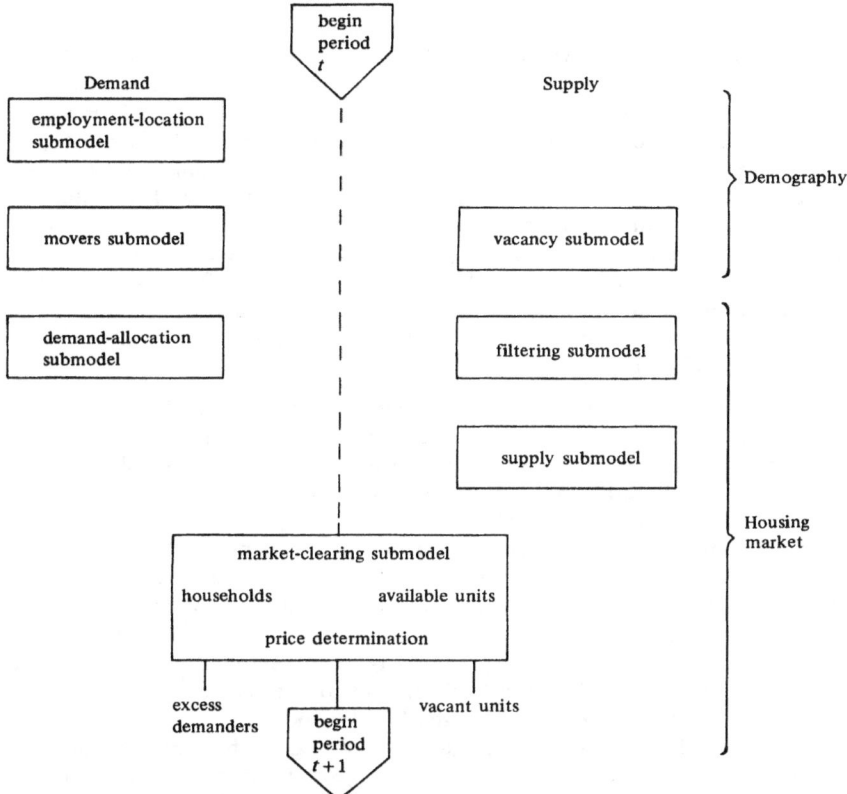

Figure 12.1. Sequence of submodels.

The demand sector. The potential demanders in each period consist of intrametropolitan movers (owing to a family size, income change or similar factors, or to a change in workplace), new households, and in-migrants, all of whom are identified by workplace and household type. They face a given price surface based on expected prices for each house type in each residential zone (expected price being obtained from historical prices via a distributed-lag function) and specified travel costs between residential and workplace zones. Intrametropolitan movers are estimated by using moving rates, stratified by household characteristics. The demand sector then allocates households, stratified by class and workplace, among available dwelling units, described by type and location. The allocations are based upon workplace and income-class-specific housing prices and travel costs. Travel costs are assumed to vary by income class and by mode, and they are the only characteristic that distinguishes locations within each housing submarket (type of dwelling unit).

The process of determining demand is undertaken in two stages. First, households select a housing type, and then they choose a location. The first stage uses econometrically-estimated demand functions, but these are interpreted probabilistically. The second stage is based upon minimization of household aggregate travel costs by use of the familiar Hitchcock-type linear-programming model. Although this separation of dimensions of residential choice has operational advantages, it contrasts with the standard NUE model in which a household chooses a housing type when it chooses a location (because density is the only housing characteristic). The other major difference is that in the NBER model the trade-off between house price, amount of housing consumption, and travel costs is implicitly analyzed via the demand functions rather than explicitly treated in the household utility function.

The supply sector. The supply sector simulates two major kinds of housing-stock adjustment: new construction, conversions, demolition, and replacement; and quality changes (filtering), assumed to take place in both an upward and a downward direction, via rehabilitation or by under-maintenance and disinvestment.

The supply model is disaggregated since the housing stock and house prices are classified by housing type and by location (residential zone). Exogenous variables include: expected house prices in the current period, all costs (construction, conversion, demolition), zoning constraints prohibiting certain types of dwelling units in some zones, amount of land and other inputs, and a forecast of demand for the current period. Subject to these data and constraints, builders perform those activities that maximize profits. This model might be solved as a profit-maximizing linear-programming problem. However, since the Detroit prototype had 33 264 possible supply activities (27 housing types × 28 possible inputs × 44 residential zones), computing and other operational costs dictated a much simpler ranking algorithm, in which supply activities were ranked from the most profitable to the least profitable, with the most profitable supply activities selected that satisfy the demand constraints. The filtering process is simulated in such a way as to represent the process whereby building owners invest in maintenance so as to equate marginal costs and marginal returns. Operationally, the model compares the change in expected price associated with a shift of a dwelling from one quality structure to another (the quality premium) with the cost of bringing about this shift through maintenance and rehabilitation (the transformation cost). If the ratio of the quality premium to the transformation cost exceeds unity, dwelling units are upgraded; if the ratio is less than unity, downgrading takes place.

The price-formation sector. As implied above, both the demand and supply sectors rely heavily on expected prices estimated by house type and location (on the demand side they allocate households to housing submarkets, whereas on the supply side they are used to calculate the profit rates for the various supply activities). There are two main methods of generating prices dynamically. One is via a dynamic price-adjustment mechanism, making use of excess demand or supply functions, to achieve convergence towards a long-run static equilibrium. The other is by optimizing techniques, using linear-programming models, Lagrange multipliers, or some similar method. As pointed out above, the NBER model adopts the second approach, specifically a linear-programming model that generates shadow prices for each housing type and residential zone. The shadow price for a given zone and housing type represents the overall travel cost that would result if one additional unit of that type was added in that zone and if households were reassigned to minimize travel costs. To avoid sudden and temporary distortions the shadow prices of several periods are used to form expected prices via use of a Koyck distributed lag. The zonal shadow prices for a given housing type are translated into location rents, via first changing their sign (they thus become travel savings) and then adding a constant to make the zone with the lowest savings in travel in the submarket have a travel saving of zero. The resulting values are the location rents for the housing submarket. Since all units in a particular submarket are assumed to be perfectly substitutable, except for location, the absolute differences in location rents correspond to one-period price differences among zones. The base one-period price for a given housing type is obtained from the zone where location rents are zero (the marginal zone), and from the minimum cost of producing that house type in that zone, assuming that suppliers in the marginal zone break even. The one-period prices for units in other zones are obtained simply by adding their location rents to this base price. The price-formation sector also determines the one-period land price in each zone. This is represented as the average of the location rents of the housing types present in the zone.

Certain compromises have to be made with theory to obtain an operational model. For example, a requirement of the linear-programming model is that the total number of households to be located must be equal to the total number of houses available. How then can the phenomena of excess supply and excess demand be dealt with? The solution is to create pseudohouseholds to live in vacant houses whereas pseudounits (perhaps temporary accommodation) absorb excess demand. Another simplification is to ignore nonwork trips in the procedure of minimization of travel cost. The assumption of the dominance of the workplace (each household knows its workplace when it chooses its home), instead of treating the choice of job and home as simultaneous utility-maximization decisions, is a further step to ease operationality. The metropolitan area is treated as virtually a closed system apart from in- and out-migration. Migration rates are assumed to be determined by the rate of a growth of employment opportunities in the area, and there is no attempt to model the rest of the world. This is not a major shortcoming since the central idea behind urban models is to develop a satisfactory explanation of the dynamics of intraurban spatial structure rather than a total world model.

To pass a few brief additional comments on each submodel, the employment-location submodel avoids the challenge of trying to develop a behavioural-location model for basic employment. Instead, its task is to transform forecasts of employment changes into spatial distributions of workers by socioeconomic and demographic characteristics. The output of the employment-location submodel is a manpower-requirements matrix— a summary cross tabulation of the total number of employees in each workplace zone for the period—to be used as input to the movers and vacancy submodels. The movers submodel estimates the number of movers by applying the relocation rates of household class to the number of households at each work zone in each household class and by adjusting the result for employment growth and decline (stratified by education, income class, and workplace). The vacancy submodel identifies the number of houses available for occupancy from the basic mobility forecast of the movers submodel.

The demand-allocation submodel allocates the housing demanders (defined in terms of workplace and household class) among the different housing types (submarkets) probabilistically. Their choice reflects household characteristics (for example, family size, high socioeconomic status) and gross house prices (that is, the market price of housing plus work-trip cost). As mentioned earlier, the filtering submodel is based upon a comparison of quality premiums with transformation (for example, improvement) costs. As a practical simplification, the filtering submodel was applied only to units available for occupancy and not to the housing stock as a whole. Also, an exogenously specified constraint of ten percent was placed on the filtering rate in any period. The supply submodel is comprehensive in that it deals with conversion to multiple-family units, and demolition and replacement in the central city as well as with new construction, which typically takes place on the metropolitan fringe. The supply submodel uses an input-output matrix which summarizes the set of efficient technologies and costs for building on vacant land or for converting existing structures. This matrix is assumed to be independent of residential zones and constant across them. As pointed out earlier, activities are ranked by profit rates, and their levels are assigned subject to the constraints of selecting the most profitable among them, as well as to the level of vacancies, zoning restrictions, and the demand forecast. The transformation costs used in the model were based on data from *The Dow Building Cost Calculator*, a manual used in the real estate, insurance, and similar industries to estimate the replacement cost of buildings. The major function of the market-clearing submodel is to locate current-period movers in the housing available. In addition it updates the trip matrix and generates a matrix of prices that form the expected prices for the next period of the model. The assignments are obtained simply by minimizing the travel costs of households in each class at a particular workplace zone for each housing type, subject to the total number of households assigned being equal to the number of movers and the number of available units. The question of substitutions between housing types is handled within the prior demand-allocation submodel; the market-clearing submodel permits only substitutions between locations.

From this description, it is easy to see how different the NBER model is from the NUE models. Although the attempt at calibration was, on the analysts' own admission, less than satisfactory, and hands had to get dirty in trying to convert the theoretical ideal into an operational and practical model that could be afforded, the NBER approach is something of a landmark in the development of urban models. It is firmly grounded in microeconomic theory, it is a simulation model in the true sense of attempting to replicate market processes but using substitute algorithms rather than the market mechanism itself, and it makes judicious use of real data by econometric estimation (particularly of housing-demand functions). The model, of course, deals in detail only with the operations of the housing market, and does not pay much attention to other types of urban activities, but then NUE models too are less than comprehensive. The NBER approach embodies several important virtues that are lacking in most NUE models. These virtues include: the use of discrete zones that permit two- rather than one-dimensional urban space; a marked relaxation of the monocentricity assumption; a highly disaggregated specification of households and dwelling units, as opposed to the homogeneous households and urban land defined solely in terms of distance from the CBD that are so characteristic of the NUE models; and a concern with adjustment paths rather than with instantaneous equilibrium. The focus on movers alone rather than total urban population recognizes that, despite high mobility rates, there are certain rigidities in the spatial distribution of urban population.

These attractions should not blind us to the limitations of the model. Some may dislike the rejection of the strict utility-maximization approach in favour of the more indirect empirical demand functions. Many would argue that reliance on the journey-to-work, cost-minimization model of residential location gives scant acknowledgment to the strides made in the theory of residential location over the last fifteen years, though the attempt to incorporate more neighbourhood quality characteristics in Pittsburgh II is a step towards recognizing environmental characteristics as a major determinant of the residential site choice. The authors admit the need for more research to be able to take more adequate account of these and other interdependencies. However, most of the compromises are reasonable and quite acceptable in a pioneering piece of research. The real attraction of the NBER model is that it shows a way out of the impasse, in which urban economists are trapped, between the arid purity of the new theoretical models and the rigid, soulless nonbehavioural mechanics of the large-scale planning models of the past.

The MIT econometric simulation model
Another urban modelling effort, that may be called the Massachusetts Institute of Technology (MIT) Econometric Simulation Model (Engle *et al.*, 1972; Engle, 1974), provides a further different approach to NUE models. It is more ambitious than the NBER model in that it attempts to specify the structural relationships that reflect the behaviour of households, business, and government, and to simulate the role both of market and of nonmarket institutions. However, judging by the sparsity of published results—with the partial exception of the macroeconomic submodel, the research was richer in promise than in execution. Its attempt at greater realism implied a need for more and better data. The stated primary purpose of the model is to permit evaluation of policy alternatives at all levels of government, from the national down to the local jurisdiction. This has certain implications for the model. Its endogenous variables would have to include social welfare criteria such as income, income distribution, access to public services, and residential segregation. Similarly its exogenous variables would have to include instrumental variables that can be changed in value to reflect a variety of policy alternatives. The model would also need substantial predictive power to aid policy evaluation.

The MIT model consists of three basic submodels: a nonspatial macroeconomic model of output, employment, and income distribution; a long-term adjustment model for population and capital stocks, including migration and investment flow equations; and a discrete intrametropolitan model of spatial allocation. The data base and the application refers to the Boston SMSA. The macroeconomic model takes the following as given: population, capital stock, technology, export demand, wages, prices, and unemployment. It generates output, employment and income distribution, and changes in wages and prices. The adjustment model provides the population and capital-stock levels for the macroeconomic model, these being derived from forecast changes in population and investment as the net effects of changes in Boston economic conditions and in national conditions. These two submodels in combination provide an aggregate metropolitan-growth model. The outputs of this are the inputs for the spatial-allocation submodel that is concerned with determining the spatial distribution of households, business firms, and the urban capital stock. Changes in population and income and in location patterns result in adjustments of the urban capital stock via construction, filtering, and demolition. Public services and tax rates have impacts on location. There are also feedbacks to the metropolitan-growth model via price effects (housing and land prices, and tax rates).

The macroeconomic model
The basic income model is a variant of the export-base hypothesis with US income as a determinant of activity in manufacturing, wholesale trade, and finance, and price effects are built in as determinants of manufacturing

and service output to allow for shifts in multipliers. The income identity is

$$y^P = M+R+W+F+S+O+G+C, \qquad (12.53)$$

where y^P is income produced, and the other variables are manufacturing, retail and wholesale trade, finance, services, other sectors (transportation, communication, utilities, mining, agriculture, etc), government, and construction respectively.

Income received is needed for many purposes. This is obtained from

$$y^R = y^R(g, y^{PB}, y^P - y^{PB}, \Pi^{US}), \qquad (12.54)$$

where
y^R is the income received,
g is the net effective tax rate,
y^{PB} is Boston earned personal income, and
Π^{US} is US profits.

The employment demands of each industry are assumed to depend on value added, real wages, and productivity, except that in the manufacturing sector the capital stock is included as an important variable. The labour-force participation rate is a function of demographic characteristics and real wages; applying this to the population (exogenous) gives an estimate of labour supply that can be related to the employment demands to obtain the unemployment rate.

Since relative price changes (in Boston vis-à-vis the United States) are assumed to affect the level of output, especially via changes in export demand for Boston's nonlocal industries, the price and wage variables have to be specified somewhat more precisely than is customary in a standard macroeconomic model. The price index is decomposed into house prices (that change in line with local land values and construction costs), the price of services (determined by wage levels), and commodity prices (that vary from the trend in US prices according to the local unemployment rate). Changes in wages are determined by changes in prices, unemployment rates, and technology.

To derive changes in the income distribution an industry–occupation matrix is used to convert employment demands into occupational demands, and then wages by occupation are used to obtain estimates of earned personal income (after adding adjustments for fringe benefits). A major difficulty in the analysis of income distribution is deriving reasonable estimates for nonlabour income.

This macromodel is static. Dynamic elements are introduced via the migration and investment equations of the adjustment model, though lagged relationships might also be introduced into the basic macromodel itself.

Long-run adjustment model
This shows how population and the manufacturing capital stock (housing and structures are handled at the individual zone level within the spatial allocation model) change over time according to economic conditions in Boston relative to the nation. Population changes as a result of natural increase and net migration. Natural increase is assumed to be a function of demographic and socioeconomic variables that determine birth and death rates. Net migration is treated as a function of the present population and of relative wages, employment, and prices. The increment to the manufacturing capital stock (investment) is assumed to be a function of the existing capital stock, lagged-manufacturing output variables, wages, prices, and tax rates in Boston compared with the US average. Also, the rate of investment at any time will be influenced by the rate of US corporate tax and the rate of interest.

The spatial-allocation model
This is the most interesting and most ambitious component of the MIT Econometric Model, but one that creates awesome data problems. The spatial submodel attempts to cope with the following problems: colocation; durable, specialized capital; the nature of land-market mechanisms; government; density and the form of urban expansion; and problems of land assembly. Colocation attempts to take account of locational interdependencies in the form of the scale and type of other activities in the same neighbourhood and access to other areas in the metropolis. The durability and specialization of capital impose severe constraints on changes in land use. Similarly, the operation of market forces in the land market is restricted by imperfect knowledge, high relocation costs, long leases, and other factors limiting the current supply of sites. The public sector influences land use, location, and spatial distribution in several ways: tax and expenditure policies; zoning, building codes, and other land-use controls; and as a major user of land (for example, for government buildings, other public facilities, and parks). A model needs to accommodate the two major types of urban growth: *intensive* growth, associated with the conversion of land use to taller buildings and higher densities, and *extensive* growth, associated with horizontal expansion of the urban periphery. Finally, lots usually become available discontinuously in specific sizes, so that land assembly becomes a complicated development problem. Areas with large tracts of vacant land can cater for much more flexible patterns of development. The aim of the MIT model is to take more adequate account of these considerations, very important in real-world cities but usually ignored both in standard theories and in operational planning models.

In the spatial submodel the metropolitan area is divided into zones corresponding to local jurisdictions, with large cities disaggregated into several zones. Land uses are classified by type according to the categories of the macroeconomic model, structures are defined in terms of type and density of use, and the housing stock is stratified by quality (low, intermediate, and high). Inputs into the model from the macro and adjustment models are the number of households and businesses which are distributed spatially, the scale of SMSA activities in the previous period, and the distribution of income. These data trigger demand and supply adjustments, as existing activities and new locators evaluate the relative attractiveness of each zone. Location and relocation take place as a result of comparison by potential demanders of user valuations (taking full account of colocational influences) with prevailing lot prices. Lagged, stochastic factors influence the locational process because of moving costs, existing disequilibrium, and expectations about future colocational configurations. There are also supply adjustments taking place, including both new construction and conversion of existing occupied sites to new uses. These adjustments depend on relative profitability, with recent past changes in land price providing a strong indicator of expected revenues. Occupancy distributions change as the net result of these demand and supply responses, and their interaction brings about adjustments in market prices. At the beginning of the period (say t) prices and available supplies are specified. These determine demand and supply, but demand intersects with initial supply to generate a price change which comes into effect at the end of the period. Meanwhile supply changes are taking place. By the end of the period there is a new supply and a new price set. These are the determinants for the market transactions at $(t+1)$.

Evaluation
The macroeconomic component of the MIT model is more highly developed than the other submodels, and much easier to implement because of the relative accessibility of the data. The reliance on export-base theory is quite understandable as a means of plugging the local economy into what is happening at the national level, but the consequential focus on short-run considerations is at odds with the long-run policy goals of metropolitan areas. The emphasis on short-run adjustments is carried over into analysis of the spatial-allocation model, and this myopia may mask understanding of the critical secular processes that can dramatically transform the visual appearance and economic and social structure of cities.

Although the spatial-allocation model makes a worthy attempt to cope with the constraints on market processes that are so critical to the evaluation of spatial structure, an attempt with which this writer is wholly sympathetic in view of the objections to NUE models stated previously (see pages 31-42), it is notable that these aims were stated prior to operationalizing the model. Compromises with optimistic objectives are inevitable given the tremendous limitations on metropolitan data, as the application of the NBER model revealed.

In his comments on a presentation of the MIT model, Muth (1972b) offered these criticisms: the lack of behavioural relationships (despite a claim to deal specifically with this problem); the concentration on the dynamics of adjustment to the exclusion of long-run comparative statics equilibrium (this is a common countercharge of alternative approaches by NUE modellers and their fellow travellers); and the model is too ambitious. Bradford (1972) shared in these criticisms, but added two more important ones. First, he reinforced the above-mentioned observation about the limited usefulness of short-run models for long-run urban analysis that results from linking local economic performance to that of the nation: "the macroeconomic model ... will at best give us a forecasting tool to tell how the Boston economy is likely to behave relative to the nation ... we shall have learned little of a convincing nature about how the Boston economy ties together, nor shall we know much about the long-term determinants of the area's wealth" (Bradford, 1972, page 101). Second, he laments the neglect of agglomeration economies that he regards as the source of urban productivity. Although these may be subsumed in the model's indices of zonal attractiveness, they need more precise specification in a model that attempts to simulate the development of the metropolitan area as a whole.

Regardless of the operationalization problems, the basic assumptions of the MIT model contrast sharply with the assumptions of the NUE model from the standpoint of representing reality. Disaggregation of the metropolitan area into zones, explicit recognition of nonresidential activity types with the constrained possibility of locating in different parts of the metropolitan area (not merely the CBD), attention to vertical as well as to horizontal expansion, a major role for the government sector, allowance for the durability of capital, severe constraints on competitive processes in the land market, some provision for locational interdependencies—these are just a few examples of how the MIT modellers attempt to face up to problems rather than ignore their existence because it simplifies the analysis. Unfortunately the gap between what is desirable conceptually and what is feasible and implementable remains almost as wide as ever.

Spatial interaction in urban models
Alan Wilson, the British geographer (and ex-nuclear physicist), has developed an alternative approach to urban models. In his recent comprehensive text on urban and regional-planning models (Wilson, 1974), he has suggested a wide range of methodologies and designs for urban-planning models. A brief summary of his most general formulation is presented below. However, although the model is intended to be comprehensive, only one segment of it is articulated in detail—the residence-workplace submodel. This is a specific illustration of Wilson's most notable contribution to urban modelling—the development of

residential-location and trip-distribution models that are based upon spatial interaction. At one level, this approach can be viewed as an extension of the standard gravity model to reflect origin and destination constraints (now known as the doubly-constrained gravity model). At another level, it can be treated as an urban application of the entropy-maximization principle. Both perspectives are grounded in probability theory. However, whereas the gravity model applies only to large populations, the entropy-maximizing model is grounded in the probabilistic behaviour of the individual unit. But since a special case (in fact, the standard model) of the entropy-maximization model, namely the equiprobable assignment case, and the doubly-constrained gravity model are virtually the same for operational purposes, the differences between the approaches are merely matters of interpretation. On occasion, the use of spatial-interaction models has been criticized for relying on averaging procedures and hence underestimating the variety of human behaviour. If this is a defect, Wilson avoids the trap by extensive disaggregation (by zone, economic activity, house type, and household characteristics).

A spatial-interaction model of residential location

The simplest spatial-interaction model of residential location (Lowry, 1964) is

$$T_{ij} = gE_j \mathrm{f}(c_{ij}) \,, \tag{12.55}$$

where
T_{ij} is the number of people living in zone i and working in zone j,
E_j is the number of jobs in zone j,
c_{ij} is the travel cost from i to j,
f is a decreasing function, and
g is a constant.

To obtain consistent results it is necessary to introduce the constraints (assuming n zones)

$$\sum_{i=1}^{n} T_{ij} = E_j \,, \tag{12.56}$$

and

$$\sum_{j=1}^{n} T_{ij} = H_i \,, \tag{12.57}$$

where H_i is the number of houses available in each zone.

If g in equation (12.55) is replaced by the multiplicative term $A_i B_j$, where A_i and B_j are calculated so as to satisfy the constraints (12.56) and (12.57), equation (12.55) can be transformed into a production-attraction (doubly) constrained model, thus

$$T_{ij} = A_i B_j H_i E_j \mathrm{f}(c_{ij}) \,, \tag{12.58}$$

where
$$A_i = \left[\sum_{j=1}^{n} B_j E_j \mathrm{f}(c_{ij})\right]^{-1},$$
and
$$B_j = \left[\sum_{i=1}^{n} A_i H_i \mathrm{f}(c_{ij})\right]^{-1}. \tag{12.59}$$

It should be noted that this doubly-constrained model takes the housing stock as given (though it could be derived separately from a supply-side model), and locates people simultaneously to residences and workplaces.

The entropy-maximizing approach focusses on individuals, and in the residential context assesses their probability of living in a particular zone and working in another particular zone (in the simple case of the one-worker household the residence and workplace will be fully specified by the journey-to-work). The assignments of individuals are the microstates, and a mesostate is the sum of a particular set of microstates, that is any T_{ij}. A macrostate is an assignment $[T_{ij}]$ that satisfies constraints (12.56) and (12.57), and a total travel-cost constraint, so that

$$\sum_i \sum_j T_{ij} c_{ij} = C. \tag{12.60}$$

Let us make the simplifying assumption that all microstates compatible with the overall constraints are equally probable. Thus the most probable mesostate is that containing the greatest number of microstates. The total number of assignments of individuals to T_{ij} is

$$S[T_{ij}] = \frac{T!}{\prod_{ij} T_{ij}!}. \tag{12.61}$$

The *most probable* assignment can be obtained by maximizing $\ln S[T_{ij}]$ subject to the constraints (12.56), (12.57), and (12.60); Wilson (1970) shows that this assignment is given by

$$T_{ij} = A_i B_j H_i E_j \exp(-\beta c_{ij}), \tag{12.62}$$

where β is the Lagrangian multiplier associated with equation (12.60)[28]. Equation (12.62) is the same as equation (12.58) except that the general travel-cost function $\mathrm{f}(c_{ij})$ has been replaced by the negative exponential function, $\exp(-\beta c_{ij})$.

This model may be extended in several ways (Senior and Wilson, 1974). The more obvious extensions include: (1) disaggregation of the housing stock by type; (2) stratification of households by income class (this implies the introduction of a housing-expenditure constraint, and this requires the spatial pattern of house prices as data input; (3) modification

---

[28] The maximum entropy of the probability distribution p_{ij} is given by
$$S = -\sum_i \sum_j p_{ij} \ln p_{ij}, \qquad \text{where} \quad p_{ij} = \frac{T_{ij}}{T}.$$

of the standard model to include dependent workers (working nonheads of households) and nonworking households; and (4) conversion for application in a dynamic model (see below).

A dynamic urban model
The structure of a dynamic general urban model as developed by Wilson (1974) is summarized in figure 12.2. It is apparent that this starts out with the familiar separation of demographic change from economic activity. These are forecast separately, partly by reference to past metropolitan performance, partly as the result of exogenous variables.

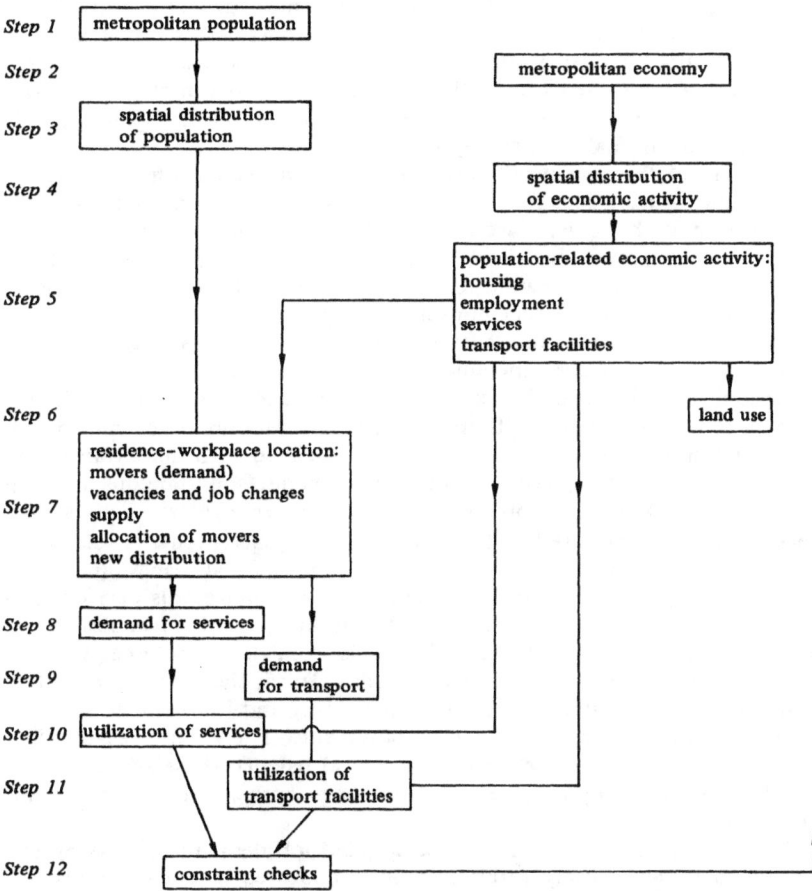

Figure 12.2. A dynamic general urban model. (Source: adapted from Wilson, 1974, page 276.)

The aggregates (people and economic activities) are then disaggregated spatially. In the economic subsystem a further important step is the determination of nonbasic activity levels (housing, services, and transport). The land-use implications of the employment forecasts are then drawn out to ensure that land-availability constraints are not violated.

The next step is the most critical one because it represents the reconciliation between population and job changes. This is the residential-workplace location submodel which allocates movers to specific workplaces and residential sites. It is here that Wilson makes his distinctive contribution He uses the principle of entropy maximization to derive variants of the doubly-constrained gravity model to obtain the most probable trip distribution between residential and workplace zones. This provides the necessary inputs for estimating the demand for transport facilities and for total services (these must then satisfy supply constraints in the utilization phases of the model—*steps 10* and *11*). Finally, the model is subject to constraint and internal consistency checks.

The detail of the residence-workplace model surpasses other attempts. For example, Wilson does not stick to the traditional assumption of one-worker households who are the sole demanders of housing. Instead, he uses four categories: working head of household, working nonhead, nonworking head, and nonworking nonhead of household. He identifies several 'transitions' that affect housing-demand pools: people changing homes (household heads are assumed to demand *houses*, whereas nonheads demand *residential places*), including out-migration from the area; conversion of nonheads of households to heads, and vice versa; workers changing jobs; entries and withdrawals from the labour force; and births, deaths, and in-migration. Several flows are associated with these transitions: housing and residential-place demand; job demand; simultaneous housing-job (and residential-place-job) demand; nonmover effects; houses released; and jobs released. The residence-workplace model is solved in a series of steps. First, the demand for houses, residential places, and jobs is estimated using the flows into housing and job-demand pools derived from the transitions described above. Second, housing and job vacancies are estimated as a result of moving, withdrawals from the workforce, and conversion of household heads to nonheads. Third, the total supply of housing and jobs is calculated by adding housing and job releases to *net* additions to the housing stock and employment. There is no reason why the demand quantities of the first step should balance these supplies. Homelessness and unemployment can be coped with via the use of dummy housing and job categories.

The fourth phase is to allocate housing and job demand by making use of the demand and supply totals previously allocated, plus exogenously-determined, transport-cost and housing-budget constraints, to solve an entropy-maximizing model. This will be similar to the model summarized in equations (12.55) to (12.62) above except that it will be much more

complicated, since it will deal with nonworker households and additional
constraints to reflect the different transitions in the demand pool and will
also be dynamic rather than cross-sectional. The fifth and final phase is
the new distribution of individuals to residences and workplaces obtained
from solving the model. It should be noted that whereas Wilson presents
a sophisticated version of the residence-workplace submodel, some of
the other submodels in his system are dealt with very simply indeed.
For instance, the aggregate demographic forecast uses the standard age-
cohort-survival technique, whereas the economic forecast is derived from
an input-output model. This is recognized by Wilson himself. His
objective is to present a very general framework within which the analyst
can vary the specification detail of the individual submodels according to
the purposes of the study and his personal tastes.

Forrester's *Urban Dynamics*
No approach to urban modelling has created more of a furore than the
systems-simulation model of Forrester (1969). For instance, Mills (1972a,
page 80) argues that "It is hard to take Forrester's model seriously as a
description of the urban economy. It contains no production functions
except input-output coefficients for land and labor. It contains no factor
prices and no factor markets that are recognizable to economists. It
contains no output prices and no output markets. There are many *ad hoc*
assumptions about lags, functional forms, and coefficient values. All are
introduced without reference to economic theory or empirical studies".
However, the major contribution of Forrester, summarized in his book
Urban Dynamics, is not so much the behavioural relationships he assumes
to exist between the key urban variables, but the methodology of urban
analysis he proposes. It is possible to disagree with Forrester's perception
of how the urban economy operates in almost every detail and yet
welcome his study because it offers a potentially useful approach very
different from those that preceded it.

Forrester denigrates earlier models of social systems because of their
assumption that the phenomenon with which they deal is assumed to be
representable by a *simple* system. This implies that cause-and-effect
associations can be identified as being proximate in space and time,
represented by simple feedback loops (usually first-order negative-feedback
loops). In fact, the real root of change may be far back in time from the
proximate apparent cause (a symptom) or may originate in some far
corner of the system. Most social systems are *complex* systems. The
structure of a complex system is not a simple feedback loop where one
system state dominates behaviour. Instead, it is a multiplicity of interacting
feedback loops, of high order (that is, there are many system states), with
internal rates of flow controlled by nonlinear relationships, and with some
of the feedbacks, that describe growth processes as well as the more
familiar corrective and equilibrating negative feedbacks, being positive.

If the city is a complex system riddled with interactions among relationships and nonlinearities, it cannot be represented by simple mathematical equations of the kind used by NUE theorists. Instead, it must be simulated on a computer. In Forrester's words (1969, page 108): "Modern mathematics deals almost exclusively with linear processes. ... Only by dealing forthrightly with the nonlinearities in systems will we begin to understand the dynamics of social behavior. ... Nonlinearity is easy to handle once we stop demanding analytical solutions for systems of equations and accept the less elegant and more empirical approach of system simulation. Acceptance of the nonlinear nature of systems shifts our attention from the futile effort to measure accurately the parameters of social systems and instead focuses attention on the far more important matter of system structure".

Complex systems have certain important characteristics:
(1) They may be counterintuitive. This follows from the difficulty of identifying the levers of change in such a system, and mistaking simple time correlation for cause and effect.
(2) Nonlinear feedback models are frequently insensitive to large parameter changes. For instance, a city may behave in more or less the same way in different institutional environments and in different historical periods. In Forrester's opinion, "social systems are dominated by natural and psychological factors that change very little" (*ibid*, page 110).
(3) The consequence of (1) and (2) is that complex systems are resistant to most policy changes.
(4) On the other hand there are a few key levers (or 'influence points') to which the system is highly sensitive, and the effects of which radiate throughout the system as a whole. Identification of these levers is not self-evident, but requires detailed study of system dynamics.
(5) Externally applied corrective action often fails because its main effect is to substitute for internal natural processes that have similar impacts (an example in Forrester's model is that job-training programmes replace upward mobility).
(6) Short-term responses are very often in the opposite direction to long-term effects. 'Worse-before-better' strategies are difficult to implement politically. Conversely, policies that appear to yield short-term benefits—and hence are politically appealing—may have disastrous long-term results.
(7) Complex social systems tend towards poor performance, aggravated by misdiagnosis, detrimental changes in design, and wrong-headed policies.

A source of dispute is Forrester's neglect of the cumulated knowledge of urban analysis. Instead of using data or the findings of empirical studies, his main source of information was "not documents, but people with practical experience in urban affairs", "the insights of those who know the urban scene firsthand", and his "own reading in the public and business press". As suggested below, the main weakness of this approach is that it was used not as a source of objective information but for policy biases and value judgments that seep into the structure of the model itself. However, one justification for neglecting available data is Forrester's argument that the barrier to progress in social systems is not lack of data but poor structural theories. Many of the important relationships cannot be measured accurately, but it is more serious to omit a relationship than to include it even at a low level of accuracy. 'A shortage of information is not a major barrier to understanding urban dynamics. ... The barrier is the lack of willingness and ability to organize the information that already exists into a structure that represents the structure of the actual system and therefore has an opportunity to behave as the real system would. When structure is properly represented, parameter values are of secondary importance. Parameter values must not be crucial because cities have much the same character and life cycle regardless of the era and the society within which they exist" (*ibid*, page 114).

Forrester's urban system consists of three sectors: population, housing, and industry. The population is broken up into three groups— managerial-professional, labour, and the underemployed. There is a housing type corresponding to each population group—premium, worker, and underemployed housing. There is one production sector, that is industry, but this is subdivided into new, mature, and declining industry. There is also a tax sector, with taxes levied on housing and industry. There is also a job sector, absorbing labour and a proportion of the underemployed group.

There are no spatial relationships in the model. Space enters only in the form of a land constraint on growth. The model makes the unrealistic assumption that the amount of land in the city is fixed. This rules out either annexation or suburbanization. The assumption obviously gives Forrester difficulties, because occasionally he argues that his system refers merely to one part of the city, for example the urban core, but in this case it is difficult to ignore the influence on the system's growth of the core's interactions with its surrounding suburbs. Some observers have suggested that the model applies to an island rather than a city, and Forrester half admits this when he argues that his model "depicts an urban area as an island in an extended environment" (*ibid*, page 249). Despite the flimsy attention paid to land, it is the land constraint that eventually stops the system from growing. Residential site sizes are fixed at one tenth of an acre, with industrial plant sizes at one fifth of an acre. Each population group has a different family size and a different housing density, so that land use for housing depends not only on the population of the city, but also upon its distribution. This is directly linked with the industrial sector because, although the labour:underemployed ratio in each business plant remains the same (2:1), the ratio of managers-professionals to labour is a function of the industry type (4:20 in new enterprise units, 2:15 in mature industry, and 1:10 in declining industry). The ability of the city to provide public services from taxation also depends upon the composition of population (premium housing has an assessed tax value six times the level of underemployed housing and double that of worker housing, whereas the public service needs of each underemployed person cost double those of a managerial–professional person and fifty percent more than those of each labour person) and on industry type (the ratio of taxes levied on new, mature, and declining plant is $5:3:1$).

The urban system is simulated over a cycle of growth, maturity, and stagnation of two hundred and fifty years. For the first hundred years the system grows more or less exponentially until the empty land is filled. There is then a progressive collapse because land exhaustion induces a decline in new construction which results in more underemployment and a deterioration in the industry mix. This has further repercussions on labour and managers–professionals. After about fifty years of decline, there is a very mild recovery and the urban system moves towards a low-level equilibrium.

Why does this happen? It follows from the structure of the system. Forrester assumes a "closed-system boundary", which implies (1) the urban system interacts with its external environment, but (2) the interaction is one-way, since the external environment influences the city and there is no significant feedback from the city to the environment (apart from out-migration, but even in this case the external environment functions merely as a 'sink'), whereas (3) the external impacts on the city are rather like random impulses that impinge upon the system but do not give it its structural growth and stability characteristics. The form of interaction of the city with its environment is one of in- and out-migration of each of the population groups (the flow of investment capital is implicit in the model, but little emphasized). Migration to and from the city depends upon how each group perceives the relative attractiveness of the city compared to the surrounding environment.

Migration of the underemployed depends on the value of the attractiveness-for-migration multiplier (AMM) which is the multiplicative influence of several factors: upward mobility of the underemployed into the labour class; housing vacancies; public expenditure; jobs; whether or not there is a low-cost housing construction programme; and a sensitivity coefficient. These are combined multiplicatively, rather than additively, for two reasons: if each variable increases, the total effect is greater than the sum of the separate effects; one factor can dominate the others if it becomes zero, thus shutting off in-migration altogether. Migration, in fact, depends upon perceived attractiveness, which in turn depends, with a lag (called perception delay), on the actual attractiveness of the city. Labour and managerial–professional arrivals are influenced by similar factors: job opportunities, housing, and tax levels; and favourable social mix (a high proportion of the in-migrant group in the urban population). Out-migration is dependent upon the converse of the determinants of in-migration; thus in- and out-migration are inversely related because they are determined by the same factors, but in the opposite direction. Each population group experiences natural increase, and there is upward mobility between groups and downward mobility from the labour to the underemployed class when jobs dry up.

Premium-housing and worker-housing are built by the private sector, whereas the supply of underemployed housing is determined by public low-cost housing programmes, downward filtering, and (negatively) slum demolitions. However, additions to the housing stock by the private sector are determined by a multiplier which is, again, the combination of several factors: the adequacy of housing (whether a shortage or a surplus); land availability; the population mix; the tax rate; the recent rate of new industrial construction; and the past rate of housing construction for the particular class.

As for industry, new enterprises are built from scratch, but then filter into mature businesses and subsequently into declining firms according to the functioning of the system. Also, declining industry is demolished, typically after thirty years, but with the length of life varying with the rate of economic growth and with land pressure. The key to the economic performance of the urban system is the rate of construction of new enterprises. This is again determined by a multiplier, the value of which depends on the supply of managers, the land fraction occupied, the availability of labour, the tax rate, and the past construction rates of new enterprises.

There is an interrelationship between tax levels and policital power. Political power is measured by the ratio of labour to underemployed. As the proportion of underemployed in the total population increases, tax assessments increase, because the underemployed have political power and a need for public expenditure, though they themselves pay very little tax.

The exponential growth in the first hundred years of the urban system is the result of the numerous positive-feedback loops built into the structure of the system. This affects in-migration and construction rates. However, the feedbacks do not remain positive forever. Size changes the conditions of the system and suppresses the positive-feedback behaviour. As the land area fills up, the construction land multipliers fall toward zero and new construction declines. This has repercussions on jobs and population, in-migration rates fall and out-migration increases until the population stabilizes. However, the rapid growth prior to year 100 first leads to some overshoot in the volume of construction. Thus, between the years 100 and 140 there is below-equilibrium depression until a stable equilibrium is reached about the year 200.

The transition of the urban system from growth to stagnation may be explained in terms of the interactions between population and economic change. In the initial growth phase the area is attractive to population, but after a certain phase the influx of people overcrowds the area, reduces the attractiveness of the city, and slows down in-migration and stimulates out-migration until population growth stops. {It is this phenomenon of exponential growth meeting a fixed capacity that leads to the familiar overshoot and collapse behaviour mode that characterizes more recent system dynamics models [such as the well-known *Limits to Growth* model of Meadows (Meadows *et al.*, 1972)]. It follows from the assumptions and structure of the model rather than from the intrinsic behaviour of cities during growth.} Stagnation-inducing economic forces result from the changing proportions of people, housing, and industry. During the rapid growth phase there is so much new construction that the decaying part of the city is relatively very small. However, as the land area is filled up, new construction can take place only as the old is demolished. In the general case, old business units and housing will dominate, and since the condition of buildings tends to determine the socioeconomic composition of the population (via labour mix, housing requirements, and tax effects), a high proportion of old buildings is associated with too high shares of declining industry and underemployed.

Forrester then goes on to consider the effectiveness of a range of alternative public and urban policies for the revitalization of cities. The results of the evaluations can be guessed at from the above description of the model. The key clue is that new industry and the associated high share of managers and professionals are good, whereas a high proportion of underemployed is bad. The programmes analyzed include: a job programme in the public sector; a training programme to upgrade the skills of the underemployed; a tax subsidy to the city from the central or regional government; and a low-cost housing programme financed from outside. These policies all fail, mainly because they attract more underemployed to the city. However, there are some urban policies that do work. Since the "problem of the stagnant urban area is one of too much housing compared to employment opportunities and too much old industry and housing compared to new ... urban revival requires demolition of slum housing and replacement with new business enterprise. Only by this shift from slum housing to new business will the internal mix become healthy" (*ibid*, page 71). Thus the solution to urban problems is to tear down slums and replace them, directly or indirectly, with new industry.

These conclusions are very interesting. In fact, in the United States in recent decades the slum-clearance–urban-renewal approach has been tried more often than the policies examined by Forrester; they have also failed, if anything, more lamentably. Apart from the adverse social effects, primarily forced relocation of inner city residents, the rate of urban revival has in most cases been much less vigorous than expected. Strategies of urban renewal have not been the salvation of the central city. Forrester's policy evaluations derive not from examination of the evidence but from the value judgments of his "sources" and from himself, and from built-in features of his model. For instance, the external environment provides an inexhaustible supply of underemployed who migrate into the city whenever housing, job, and other conditions are favourable. Attempts to improve the situation of the underemployed tend to attract more underemployed into the city. This leaves the lowest strata of society as badly off as before. Moreover, the larger numbers of underemployed consume more land for housing, leaving less for industrial plants. Another problem with the model is that there are no constraints on the growth of industry in the model. In particular the model assumes no limitations on the demand for industrial output. As a result, measures to stimulate industry always work.

One of the value judgments underlying the model, and carried over into the model itself, is that urban problems cannot be solved by outside help. The relevant assumption in the model is that interaction between the city and its outside environment does not affect the structure of the urban system or alter its dynamics. The city is "a living, self-regulated system which generates its own evolution through time. It is not a victim of outside circumstances but of its own internal practices. As long as present practices continue, infusion of outside money can produce only fleeting benefit, if any" (*ibid*, page 129). What practices would be effective? Expanding the proposal discussed above, the "city, by influencing the type and availability of housing, can delay an increase in the immigration rate until internal balance is reestablished. The city must press for removal of aging housing before deterioration creates an imbalance in the urban system. Because aging is continuous, the renewal process must be continuous instead of occurring in waves several decades apart. At the same time industrial parks should be established within present decayed residential areas to generate jobs for those already living in the city. Favourable city regulations and new tax policies should be designed to attract the kinds of industry most needed for revival" (*ibid*, page 128). It hardly needs spelling out that this prescription is a green light for business.

Despite the severe criticisms of urban dynamics, especially by economists, there are many advantages to the approach that have been swept away and ignored in the deluge of controversy. First, in a field where there is a chronic shortage of data, and most of the testable models are very simple in structure, the simulation model offers a way of making more rapid progress. Although economists dislike the simulation approach generally, perhaps because it uses tools and a methodology with which they feel uncomfortable, a little reflection suggests that what is wrong with the Forrester model is the structure he put into it rather than the methodology *per se*. The structure of the system reflected the results he wanted to obtain from the model, perhaps subconsciously. The 'overshoot-collapse behaviour mode' is characteristic of the series of MIT models that derive from industrial dynamics, but it is not an inevitable property. Although the concept of positive feedback loops has not been used much by economists (W J Baumol and Gunnar Myrdal are notable exceptions), presumably because the equilibrium solutions of neoclassical theory rest exclusively on the hypothesis of negative feedbacks, it has considerable potential in many areas, of which urban analysis is merely one.

The scientific merit of the system-dynamics approach has been masked by the casual way that Forrester used to obtain his model input. Even if data are scarce, there is a sufficient body of knowledge about how urban economies function to avoid the sole reliance on the analyst's own prejudices and on what businessmen and city managers tell him. As implied above, this is not a short-cut method of obtaining a deep understanding of what makes cities tick; rather, it tells us far more what makes city businessmen tick. Ignoring available empirical evidence leads to errors in the model's structure. A simple example may illustrate this point. Forrester assumes that the determinants of in-migration also determine out-migration. If the abundance of some variable stimulates in-migration, its scarcity encourages people to move out. Thus in- and out-migration are inversely correlated. Examination of the evidence makes it apparent that gross and net in-migration per capita tend to be positively correlated (Cordey-Hayes, 1974). Attractive cities have high out-migration and very high in-migration rates, whereas declining cities have low out-migration per capita rates and net migration rates close to zero. This finding suggests the need for an interdependent interurban model rather than one which reduces the world to one city and its external environment.

The approach has other virtues that, although not exclusive to system dynamics, contrast sharply with standard NUE models. The focus on secular growth paths can provide quite different insights from the short-run equilibrium models. Although the time horizon of two hundred and fifty years would appear too long to assume that the urban system retains the same structure, this flaw is not as serious as it seems at first sight, in view of the model's insensitivity to parameter changes. For instance, if the 'periods' in the model were assumed to be less than one year, this would simply mean that the values for some parameters would be too high. Some economists have criticized the model for not including prices. It is arguable that price changes are much less important in accounting for long-term variations in growth rates. The emphasis upon migration changes as generators of growth and stagnation gives the model a relevance that is not found in the standard 'closed-economy' models, even if the details of the migration submodel are challengeable.

This general support for studying secular growth does not imply endorsement for Forrester's description of the character of this growth. The determinants of growth are expressed in a very simple form, perhaps because of limited understanding of the nature of urban growth processes, but the simple structure is very specific. The rate of change of the variable depends, *inter alia*, on an earlier value of that variable, which is itself undergoing change. For instance, the in-migration rate of the underemployed depends on the existing level of the underemployed, which reflects part in-migration rates. This type of difference equation has an exponential solution. By framing the migration equations in this way, Forrester in effect assumes that urban growth is exponential. But this is an assumption not an observation. There is no evidence that urban growth is exponential even in the first hundred years of growth. Moreover, as stated above, growth collapses in the model only because of the two unrealistic assumptions: exponential growth and a fixed amount of land. Since neither assumption is valid, the explanation of collapse in the urban system is fraudulent.

At bottom, Forrester's criterion for evaluating the effectiveness of a policy is how it affects the ratio of underemployed to professionals in the population. The results he obtains are very much influenced by the particular structure of his model and by his choice of the size of the planning region. If, for example, the in-migration of underemployed depends not linearly on the level of underemployed plus the labour group but on the square root of the labour group, that is, substituting $E_t^{0.5}$ for $(E_t^u + E_t)$, the equilibrium level of the underemployed (approximately two hundred years out) is much lower than in the original formulation. Also, if policies of the type rejected by Forrester—such as job training— are applied nationally within a system of cities rather than in the isolated city, they may be effective in reducing underemployment (Kadanoff, 1971; 1972). However, the implication that a city's problems might be amenable to national policies promoted by the central government is anathema to the value judgments of Forrester and his advisor-sources.

This last point raises a paradox in Forrester's model and in his justification of it. His book lays considerable stress on the need to treat social systems as complex systems and on the drawbacks of simple systems. Yet he assumes that the city is in effect a closed system: first, because the city interacts only with an external environment that spews in or sucks out migrants (and capital) and has no other function, whereas the growth of a city can be understood only within the context of a fully interdependent system of cities; second, because the urban economy has a closed boundary. The result is that Forrester's city is a simple system after all. There is some fancy footwork involved in demonstrating the overshoot–collapse behaviour mode that almost always follows because of the many interdependencies in the system, but once this has worked itself out the system settles down on an equilibrium path not very different from the neoclassical stationary state. A complex system would require the city to be treated as an open system in a nontrivial way. The neglect of interurban competition, the assumption of no demand restraints, and the consequences of assuming unlimited resources from the environment (apart from land) are merely three out of many adverse consequences of Forrester's limited perspective of what constitutes an open system.

Finally, the model is eventually aspatial, since it consists only of one area—the city, and occasionally merely one part of the city—and the external environment, which is not modelled. A satisfactory urban-simulation model would need to deal with intraurban spatial relationships, and this would require disaggregating the city into a sizeable number of discrete zones or, as in NUE models, dealing with smooth density and rent gradients. There is no reason why the model could not be disaggregated, apart from capacity constraints on the computer. Presumably, in this case, the cycle of growth, collapse, and stagnation would vary sequentially, with inner-city zones collapsing sooner than zones more distant from the centre. The performance of the urban system might then have similarities with Blumenfeld's concept (1954) of the crest of the wave of metropolitan expansion and with diffusion wave analysis (Morrill, 1968; 1970).

Political economy

1 Marxism and the city

For those who are jaded with the abstractions and the conventions of the neoclassical city, which treat the allocation of urban land as the outcome of an orderly competitive process analogous to the rules of a mid-afternoon parlour game, turning to the city as seen through the eyes of the radical political economist is a welcome relief. Even for those who accept nothing of this alternative vision, it at least forces them to face their most cherished assumptions with new challenges. Moreover, it is the vision which is different, not necessarily the reality. Many liberals and Marxists share common ground in their description of urban ills, failures, and maldistributions. Where they differ is in their diagnosis and their prescriptions. They recognize the same problems, but from a different perspective.

To this extent, therefore, a brief look at what Marxists have to say about the city is not a digression, but a slight twist to the main theme. Moreover, regardless of one's attitude to the specifics of the diagnosis, the Marxist interpretation has certain general virtues which are appealing, and which cannot be found in the equally ideological neoclassical literature. For instance, cities have developed in real historical time. The ahistorical abstraction of traditional theory ignores this critical fact, and as a result many of its predictions are invalid. The crucial differences in the United States between old, eastern cities and newer, western cities—so clearly documented by Moses and Williamson (1967)—are a clear illustration of this point. No one could criticize Marxist theory for ignoring the historical context, even if one might not agree with the particular brand of historiography.

Another plus for the radical analysis is its recognition of the importance of social and political forces in urban life, even if they are assumed subservient to the yoke of the mode of production. Furthermore, in their insistence on the axiom that it is production rather than distribution which really matters, the Marxists have paradoxically focussed on the distributive consequences of urban allocation and growth which, in the opinion of this writer among others, are so critical. Perhaps even more important is the argument that urban problems are not problems in a vacuum but reflect society and the economy at large. This has also been accepted by the more perceptive of non-Marxian urban analysts (such as Alonso and Webber) and summed up in the epigram: "Urban problems are problems *in* cities not *of* cities".

Before discussing the radical views of the city, it is only fair to state my own position. Much of the radical assessment of urban problems is, at least, partially convincing in the sense that urban spatial form and its changing structure has been highly profitable to capitalism, and may have been stimulated by capitalism. The prescriptions are much less satisfactory, however. Marx and Engels argued that large cities were the product of capitalism and the main source of the contradiction between urban and rural life. As Engels put it: "in the huge towns civilisation has bequeathed us a heritage which it will take much time and trouble to get rid of. But it must and will be got rid of, however protracted a process it may be ... the great towns will perish". There are at least three drawbacks with this prediction. First, in those countries already urbanized with multimillion metropolises, it is difficult to imagine any chain of events—short of earthquakes and nuclear holocausts—that would reduce these cities to dust. Second, to many people—including myself—their destruction would be undesirable. Cities may create pollution, congestion, crime, poverty, and misery, but they also reflect a quality of life, a culture, and a sense of anonymous camaraderie that cannot be found elsewhere. This is not to scorn the virtues, attractions, or cultural potential of small-town or rural life, but does imply that these are different. If I really believed that the social costs of big cities were intractable without dismantling the city itself, perhaps I would be less zealous for its preservation. For me, a capitalist society with big cities may be far from perfect, but a socialist society without big cities would be less than idyllic. Third, combing the socialist literature for predictions of the ideal urban structure is very unprofitable. For instance, the proposals of Gutnov *et al.* (1970) for cities based on multiples of New Units of Settlement (NUSs), each with a maximum of 100000, a civic centre surrounded by residences and with peripherally-located schools and sports facilities, cloaked in greenery, with pedestrian access, and a rapid transit system to out-of-town industrial sites, sounds more like the pipe dreams of the more starry-eyed planners of the British New Towns than the socialist dream, not to mention the socialist reality. They might be described by that most disparaging of Marxian epithets as 'utopian'.

Cities and capitalism
The city cannot be understood in isolation from the economic system as a whole. On the contrary, there is a clear relationship between urbanism as a *social form*, the city as a *built form*, and the *dominant mode of production* (Harvey, 1973, page 203). Indeed, cities exist because of the geographical concentration of social surplus product and should be regarded as a device for creating, extracting, and concentrating surplus value. The growth of the city involves the transfer of profits and investment from the primary circuit of production into the secondary circuit of financial speculation and the development of real estate. The latter may be 'socially necessary' but it is 'unproductive'. This has been a strategy for stabilizing capitalism by stimulating new investment outlets, but in the long run it is contradictory. "What has to be explained, however, is how returns can be higher on the secondary circuit over any

length of time ... If all capital chases rent and no capital goes into production, then no value will be produced out of which the transfer payment that rent represents can come" (Harvey, 1974, page 241). The secondary circuit (see below) is more crisis-prone. Moreover, whereas in the past the city absorbed much of the surplus value in forms of conspicuous consumption (monumental buildings), the interconnections between urbanization and industrial development are now much stronger than ever because of the necessity to stimulate consumption, and one of the most profitable ways of doing this is by promoting suburbanization. If the financial support for suburbanization collapses, the associated consumer durable industries will fall with it. Moreover, many large metropolitan cities in developed countries derive much of their surplus value from the production of use values in the rest of the world (the developing countries); this makes them vulnerable to socialist revolution elsewhere.

An internal difference in interpretation has developed within Marxist thought about these interconnections. Lefebvre (1970; 1972) argues that urbanism has replaced industrialism as the dominant characteristic of advanced capitalism. He draws a distinction between two circuits in the circulation of surplus value: the primary circuit, that is, industrial activity involving the conversion of raw materials and natural resources into objects of utility; and the secondary circuit, the creation and extraction of surplus value out of speculation in property rights and in the use of distributed profits in such spheres as the housing market. As capitalism advances, "whereas the proportion of global surplus value formed and realized in industry declines, the proportion realized in speculation and in construction and real estate development grows. The secondary circuit comes to supplant the principal circuit" (Lefebvre, 1970, page 212). Harvey (1973) has criticized this view on the ground that although it may indicate a trend it is not yet established that industrialism has been replaced by urbanism as the key trait of capitalism. The role of urbanization in the composition of investment has been created by the dynamics of industrial capitalism, the stimulation of consumption in cities to maintain effective demand is due to capitalism, and the production, expropriation, and circulation of surplus value can still best be described in terms of the operation of the industrial system. As yet, Lefebvre's claim that "urbanism now dominates industrial society" is not entirely convincing.

Rent
Traditional neoclassical theory treats rent as a return to a scarce factor of production, analogous to interest on capital or the wages of labour. The Marxists argue that this is a misconception. Instead, "rent is, in effect, a transfer payment realized through the monopoly power over land and resources conferred by the institution of private property" (Harvey, 1974, page 240). Perhaps certain pieces of land, say downtown sites, are more productive than others, but this is not a consequence of nature so much

as the construction of the man-made resource system of the metropolitan city. If such land is scarce, it is an artificial scarcity in the sense that the permanent fixed assets of the city centre are highly localized.

Rent arises from "the monopoly by certain persons over definite portions of the globe, as exclusive spheres of their private will to the exclusion of others" (Marx, 1967 edition, volume 3, page 615). The most striking illustration of this exercise of monopoly power is absolute rent, and this is seen most clearly in the operations of the residential housing market in the central city. The opportunities for extracting absolute rent are increased by the fact that in any big city there is a subjugated class of housing consumers with no chance of obtaining mortgage credit and with no choice but to rent accommodation wherever they can get it at the price charged. There is a class of landlords willing to cater for this market provided that the rents are high enough. However, they have class power because they can afford to withdraw some of the units under their control from the market if the rate of return is not high enough. If the rate of return is below that earned on the capital market, they will reduce maintenance and actually disinvest. With declining maintenance, the housing stock declines in quality and the worst units are taken out of the market. The reduction in supply forces up rents because the potential renters have nowhere else to go. The rate of return drifts back up to the threshold level or beyond. These higher rents are extracted only from the poor. "The rich, who have plenty of economic choice, are more able to escape such consequences of monopoly, than are the poor whose choices are exceedingly limited. We therefore arrive at the fundamental conclusion that the rich can command space whereas the poor are trapped in it" (Harvey, 1973, page 171).

The opportunities for extraction of absolute rent are increased by the effects of residential segregation, which breaks up the aggregate metropolitan housing market into a relatively noncompeting set of islands in which scarcity can be created. The situation is aggravated by the activities of local governments (for example, zoning practices) and by the operations of financial institutions (for example, severe restrictions on the type of properties and class of households which can gain access to mortgage credit) that squat on the top of the exploitative hierarchy.

However, even the middle classes are not immune from exploitation—though by the collaborator of the landlord, the speculator-developer. What happens on the suburban fringe is intimately connected with changes in the central city: redevelopment displaces housing; deterioration of poor neighbourhoods due to withdrawal of housing by landlords scares residents of adjacent neighbourhoods into suburbanization; and so on. The activities of speculator-developers are supported by existing institutions. Governments reduce uncertainty in land-use competition by zoning and provision of public infrastructure, whereas the central government stimulates speculative development via favourable tax incentives.

Despite their damning critique of neoclassical rent theory as "a disconcerting mixture of *status quo* apologetics and counterrevolutionary obfuscation" (Harvey, 1973, page 193), the radicals sometimes recognize that out of the neoclassical wreckage it is possible to salvage a concept of rent that can offer guidelines to socially just and beneficial decisions concerning land use, perhaps even for revolutionary action. Such a concept might reflect rent as a shadow price of social choices foregone (Harvey) or as a measure of the *net* social benefits of urban agglomeration (Edel, 1972).

Conflict theories of land-use competition

In three recent papers Scott (1975a; 1975b; 1976a) has challenged the idea that land-use competition generates a mutually satisfactory equilibrium by replacing the neoclassical model by what he calls a 'classical' model, extending a tradition that can be traced back to Ricardo and Marx. In a static framework the benefits of urban agglomeration are to a large extent fixed, and must be shared among three groups: workers (suppliers of labour); capitalists (demanders of labour); and landlords, who gain from the absolute scarcity of urban land (a form of absolute rent) and who skim off spatial differences in benefits in the form of differential rent. Since these groups have to be satisfied out of a finite output, they must be in conflict. Although landlords are in competition with each other they may act as a group to withdraw land from the market if the expected rate of return falls below the desired level (see Harvey, 1973). Similarly it is in the interests of capitalists to keep wages as low as possible so as to maximize profits. Scott draws a sharp distinction between the central wage rate (the gross wage paid by CBD employers) and the residual wage, the net income of workers after payment of transport costs and rents. Improvements in urban productivity help landlords more than any other class because they increase workers' competition for urban space, a phenomenon that allows landlords to increase rents (Edel, 1972). Even a subsidy to public transit may yield little more than a transient benefit to workers, since lower commuting costs can so easily lead to higher rents and/or lower wages. In the latter case, two forces are in operation—the increase in the residual wage and the spatial extension of the labour market. Whether this view of urban competition is more or less valid than that of the neoclassicists can be resolved only by reference to subjective value judgments and cannot be settled by empirical testing.

In his third paper, Scott (1976a) develops these ideas in a simpler but more formal manner by showing the interdependence between Sraffa's model of the production process and the von Thünen model of land rent. Although the analysis is framed in terms of agricultural crops (von Thünen's famous concentric rings), the extension to urban land use is quite straightforward. The Sraffa nonspatial model can be expressed as a system of simultaneous equations

$$\left(\sum_j q_{ji} z_j\right)(1+\pi) + E_i w = z_i ,\qquad(13.1)$$

where
q_{ji} is the quantity of output j used in production of one unit of i,
z_j, z_i are unit prices,
E_i is the amount of labour used in production of one unit of i,
π is the endogenously determined rate of profit, and
w is the wage rate, paid at the end of the production period, the size of which is determined relative to π as the outcome of an economic and political power struggle between labour and capital.

To convert the model into spatial terms, a new price level, v_j, is substituted for z_j in the above equation, where v_j is the delivered price at a central market, and includes transport costs and rent (though $v_i = z_i$, that is, intermediate inputs from a sector to itself are priced at their production price net of transport costs and rent). More specifically,

$$v_j = z_j + t_j(\bar{j}) + p_j^d(\bar{j}) + p_j^s ,\qquad(13.2)$$

where
$t_j(\bar{j})$ is the transport cost of one unit of j from the outer boundary, $\bar{j}$, of its production area,
$p_j^d(\bar{j})$ is the differential land rent (determined by savings in transport cost) per unit of output at its production boundary (for the marginal land use k, $p_k^d(\bar{k}) = 0$), and
p_j^s is the scarcity rent per unit of output arising from the shortage of land relative to the demand for its products (that is, k is fixed and finite).

If there is no boundary constraint on the area of cultivation (or, in a city, on the area of urban land uses), there is no scarcity rent, and $p_j^s = 0$. Scarcity rent is similar to the Marxian concept of absolute rent, a levy imposed by the scarcity of land used in commodity production.

This modification shows the interdependence of the production process as analyzed by Sraffa and the spatial system of von Thünen. Viewed from a particular perspective, this model represents the spatial system as the arena in which a tripartite conflict between labour, capitalists, and landlords is fought. The implications of this perspective have hardly been touched upon by urban analysts, but could have reverberations in the diagnosis and interpretation of the whole range of urban problems.

On similar lines, a neo-Marxian model of the functional distribution of income and of growth in an urban setting has been developed by Farhi (1973)[29], and is achieved by combining an aggregate growth model of the Solow type (Solow, 1956) with an urban model similar to that of

---

[29] Farhi's elegant mathematical model merits careful study. For reasons of space only a brief verbal description is presented here.

Alonso (1964). The amendments include: the assumption of increasing returns to scale in production on the ground that this explains agglomeration in cities rather than evenly dispersed production; a threefold division of output into profits, wages, and *absolute* rent (accruing to landowners as a result of the scarcity of urban sites); housing and transportation expenditures and rent are assumed to be increasing concave functions of city size, and inject decreasing returns into the model.

If we assume for simplification that utility depends on consumption alone, locational equilibrium will depend upon equal consumption everywhere. By using average values we obtain

$$u = \bar{c} = \frac{Y - \Pi - P^H - T - P}{E} , \qquad (13.3)$$

where
$\bar{c}$ is the average consumption,
Y is total output,
Π is profit,
P^H is housing expenditure,
T is transport expenditure,
P is absolute rent, and
E is the labour force, assumed equal to city size.

The effect of the agglomeration economy is represented by the term $(Y - \Pi/E)$ which shows how wages increase with city size owing to scale economies in production, whereas the remaining term $-(P^H + T + P)/E$ reflects the diseconomies of scale (congestion effects). Utility is maximized when the increase in wages with city size is exactly offset by the increased costs of urban living (assuming that the price of consumption goods, other than housing, land, and transportation are invariant with city size), that is when

$$w' + h' + t' + p' = 0 , \qquad (13.4)$$

where $w' > 0$, but $h', t', p' < 0$ ($w = Y - \Pi/E$; $h = P^H/E$; $t = T/E$; and $p = P/E$). The wage effect is determined by the extent of economies of scale in production, and by the outcome of capitalist-worker bargaining power; this contrasts with the neoclassical model in which the functional distribution of income depends on the assumption of full employment and on conditions of marginal productivity. Housing and transportation costs increase with city size because of congestion (construction costs are a function of the density of development, transportation costs are a function of average distance travelled and the degree of traffic congestion). Absolute rent arises because of the monopoly power of landowners, and increases with the competition for urban space.

An important implication of the analysis is that total output is determined by production conditions (as given by the production function), but the distribution of output is determined by conflicts among capitalists, workers, and landowners. In the growth model developed by Farhi, the three key distributional parameters are $\bar{\bar{u}}$ (equilibrium utility level of workers), Π (the rate of profit), and $p_{\bar{r}}$ (the absolute rent *per household* measured at the urban boundary $\bar{r}$). He shows that when one of these parameters is held constant there is an inverse relationship between the other two. There is even severe conflict, rather than collusion, between industrial capitalists and landowners. The only way to avoid zero utility is to increase the capital–labour ratio indefinitely, and this implies a zero rate of profit. An attack on land monopolies by industrial capitalists is a means of staving off the rate of decline in the profit rate, but not indefinitely.

The role of finance capital

Much of the Marxist analysis of the strength and resilience, but also the contradictions and instability, of advanced capitalism stresses the unholy alliance between monopoly and finance capital: "the ultimate power to organize the production and realization of value in society lies in the hands of finance capital" as a result of a necessary "inner transformation of capitalism ... finance capital comes to exercise a hegemonic power over industrial production as well as all other aspects of life" (Harvey, 1974, page 252). This control and its effects can be illustrated *inter alia* at the urban level: "The perpetual tendency to try to realise value without producing it is, in fact, the central contradiction of the finance form of capitalism. And the tangible manifestations of this central contradiction are writ large in the urban landscapes of the advanced capitalist nations" (*ibid*, page 254).

Although urban infrastructure accounts for a substantial share of investment and gross domestic product in capitalist economies (generalizing broadly, about one-fifth to one-quarter of GDP is fixed investment of which perhaps one-half to two-thirds is absorbed by the built environment), its share is falling. In the United States and many Western European countries, much of the urban capital stock was built before 1920, and since then there has been a shift from urban infrastructure (housing, other buildings, utilities, transport facilities, etc) to producers' durables and to consumers' durables (such as automobiles and household appliances). Yet urban infrastructure retains an importance far greater than implied by its direct capital requirements. Many other activities, such as transportation and the production of household appliances, are closely complementary. Its long life-span and its financing, via long-term debt, links urban infrastructure to the financial sector. Furthermore, the scale, distribution, and quality of urban infrastructure may influence the overall performance of the economy. For instance, if congestion costs are high as a result of inefficiencies in the transportation system then (in Marxist terminology) the rate of exploitation of labour power must be increased if profit rates are to remain constant. This is consistent with Farhi's (1973) contention that transport costs, congestion costs, and other diseconomies of urban scale provide another reason for the falling rate of profit under capitalism, in addition to those suggested by Marx.

The Marxist thesis, as applied to the United States economy though it is almost as relevant to other developed countries, is that governments reacted to the effects of the post-1929 depression by stimulating the expansion of financial institutions (the financial superstructure) with the express task of stimulating effective demand via the provision of credit for purchase of houses, cars, and other consumer durables. Thus the character of urbanization since that time—in particular, the growth of a high-consumption, individualistic, and suburban life-style—has been directly moulded by the financial superstructure. Furthermore most of the urban problems of contemporary America—racial and class segregation, neighbourhood decay, speculative development, differentials in local accessibility to good quality education, health and other social services, central-city–suburban fiscal conflicts—can be traced, or at least linked, to residential differentiation in cities, and this is to a large extent a consequence of the manner in which finance is channeled into urban housing markets. In these ways there is a direct interdependence between aggregate policies to promote effective demand in the national economy and the dynamics of urban spatial structure.

These developments lead to at least two major contradictions. First, speculative suburban development to boost profitability is successful only at the expense of the central city, in which financial institutions are directly involved as holders of long-term debt. Second, the suburbanization associated with shaping the city as a consumption system conflicts with its efficiency as a production unit. Thus "urbanization manifests and perhaps contributes to many of the contradictions implicit in a dynamic capitalist mode of production" (Harvey, 1975, page 45).

Suburbanization

The suburbanization phenomenon and its interpretation provide a useful illustration of the differences between Marxist and traditional urban theories. To the neoclassicist the development of suburbs has resulted from the expression of household preferences—the outcome of utility-maximizing decisions associated with rising incomes and changing tastes and life-styles. To some planners this process has led to certain wasteful side effects in terms of high infrastructure costs, sprawl, and heavy investment in roads, but these can be corrected by appropriate remedial action by planning decisions and policy measures. To the Marxist suburbanization and monopoly capitalism are directly connected: "it is consumerism, not affluence *per se*, that has encouraged suburbanization, and ... consumerism and monopoly capitalism are functionally related" (Sawers, 1975, page 16).

The argument is that the suburbs developed as a result of actions taken by capitalists to maintain the rate of profit. This objective required new methods of production (the land-intensive assembly line), the creation of new outlets for effective demand (cars, suburban houses, consumer durables), and a switch of profits from the primary circuit of production to the secondary circuit of speculation in property rights. These interrelated changes resulted in the urban spatial structure associated with monopoly capitalism: the decentralized metropolitan area with its striking contrast between affluent suburbs and run-down central city.

The analysis is strengthened by arguing that suburbanization is not a recent trend but can be directly connected with capitalist expansion in the late nineteenth and early twentieth centuries, though reinforced by the search for new investment opportunities in the period following the 1929 depression. Even a century ago Engels wrote that "everyone in a position to do so prefers to live in the suburbs rather than in the centre of a smoky town" and painted a picture of concentric zonal land use with the rich living on the outskirts served by good transport facilities and insulated from the poverty in the central city that complemented their affluence, which provides at least as plausible a vision of the contemporary American city as that of more recent concentric-zone theorists such as Park and Burgess (Harvey, 1973, pages 130-136). Sawers asserted that the American auto industry, especially General Motors, destroyed the US urban trolley-car system by buying up, through subsidiaries, more than one hundred trolley systems in forty-five cities, replacing them first with buses and increasingly by the much more profitable private cars[30]. This strategy of the 1930s and early 1940s was closely associated with the growing dominance of the auto industry in the US economy, measured by its direct and indirect contribution of 15-20 percent to GNP and the fact that the major auto and related firms account for up to 25 percent of all corporate profits (in good years). Thus "the auto and monopoly capitalism have become so closely identified that it is difficult even to imagine an auto-free developed economy" (Sawers, 1975, page 11).

Similarly, urban relocation of industry to the suburbs was motivated—according to this analysis—by the profit potential to be gained from substituting capital- and land-intensive techniques (par excellence, the one-storey assembly line factory) that alienate and reduce the marketable skills of the workers. Even more important is the manipulation of consumer tastes to create and sustain the demand for new 'necessities'—the house, the car, the dishwasher, and the other trappings of suburban lifestyles. "This consumerism and privatization of life under capitalism reach their quintessential expression in the modern American suburb. Every family owns thousands of dollars worth of appliances that it uses only

---

[30] In the case of Los Angeles, Hilton (1976) has convincingly challenged the "villain theory" showing that neither General Motors nor any of its subsidiaries controlled the Pacific Electric. Also, nationwide the trolley car systems were unprofitable and failing even before they were taken over.

minutes a day. Each family has its own lawn mower, its own indoor and outdoor recreation space, its own automobiles, and its own children. A castle in the suburbs, surrounded by a moat of grass, is the nightly destination of millions driving home through the rush-hour in the absolute privacy of their very own motor car" (*ibid*, page 17). This did not happen accidentally by the spontaneous evolution of consumer sovereignty but results from the fact that in a capitalist society households have nothing else to do but consume.

These processes were cumulative and took a long time. These facts are important in explaining how the processes occurred. Firms move to the suburbs, and households follow to be closer to work. As more households suburbanize, more firms decentralize to be closer to labour markets and to consumers. The extension of auto ownership makes more suburbanization feasible. More suburbanization leads to the demand for more roads, which encourages more people to move out of the central city. As the auto industry grows, it adds its pressure for more roads. The process is self-generating and cumulative, and all its different aspects are linked to the mode of production—capitalism.

Suburbanization is a type of filtering in the sense that historically the wealthy have moved out first. The income stratification of society, reflecting the class stratification, has permitted the suburbanization of cities to take place almost imperceptibly over three generations. Had this taken place in a very brief interval of time, its irrationality would have been obvious. In prevailing conditions the relocation decision of the individual household made sense, but the result was "millions of individually rational decisions adding up to a socially irrational one" (*ibid*, page 18).

Although the pattern of urbanization has served to bolster capitalism by providing it with new opportunities for creating surplus value, suburbanization has also brought new contradictions that increase the instability of the mode of production. The ruling class, operating from headquarter offices of the central city (the exploitative, corporate city of Edel, 1972), requires a bureaucracy of managers, professionals, and technocrats who, because of the deterioration of the central city, live at increasing distances from their place of work. The deterioration of the central city has been aggravated by the operations of the financial superstructure which has had to walk a dangerous tightrope between writing off the value of capital assets (in the central city) in order to stimulate development and effective demand (in the suburbs), and preserving the value of debts incurred by financial institutions (especially in the inner city). The flight to the suburbs has been induced to stimulate demand, but it also rocks the local government and national financial structure.

Another consequential contradiction is between the city "as a consumption artifact" (Harvey), the creation of monopoly capitalism, and the city as a production unit, the creation of competitive capitalism. Liberal attempts to resolve this contraction, such as urban renewal to restore the livability of the central city for the wealthy or the development of mass urban transit, have failed miserably. The defense mechanisms of the middle class to insulate themselves from these increasing inefficiencies and their concomitant fiscal problems of the central city, such as resistance to metropolitan annexation and exclusionary zoning to keep out the poor, have hitherto been very successful—but only at the expense of intensifying the contradictions. To resolve the contradictions successfully, for instance by changing city land uses so as to create an improved spatial relationship between workplaces and residences, requires "not only containing those processes whereby an effective demand is currently stimulated but challenging the exclusionary powers of suburban jurisdictions—in other words, challenging the myths of political rights and consumer sovereignty" (Harvey, 1975, page 43). Since the government cannot be relied upon to effect these changes, since "the state is nothing more than the organized collective power of the possessing classes, the landowners and the capitalists" (Engels, 1970 edition, page 65), the only solution is to change the mode of production. Liberal solutions must fail because, although they acknowledge the problems, they try to correct them within the existing societal framework, that is, they "want to maintain the basis of all the evils of present-day society and at the same time ... want to abolish the evils themselves" (*ibid*, page 41).

Suburbanization also reinforces class, race, and sex antagonisms which increase the instability of modern urban society. According to one ideal scenario, urbanization might have been expected to erode class consciousness as face-to-face urban services replace the impersonality of man-machine production and as the harmony of community interests replace the class conflicts that revolve around ownership of the means of production. In fact, in capitalist cities such as those found in the United States the opposite has happened. These conflicts have been underlined and reinforced by the rigid spatial segregation of classes epitomized in the central-city-suburb dichotomy. This has been accentuated by racial segregation. As a result, instead of community solidarity being a salve to the economic class struggle, intercommunity conflicts of interest have replicated it. The competition, rivalry, and hostility between central-city and suburban communities have benefited only the central-city landlords, the redevelopers, and the suburban speculators.

The sexist implications of suburbanization are a little different in that conflicts have emerged within more than between social classes. The middle-class suburban housewife, half-free from the drudgery of housework and child-rearing by household appliances and the oral contraceptive, is more enslaved than ever in long solitary confinement from her commuter husband in the mindless pleasure palace of the suburban home. Meanwhile, the woman in the central city is the most oppressed member of the industrial reserve army, exploited in the factory and the shop, exploited in the home, and exploited on the streets of the ghetto. The hysterical demand for women's rights among a small, but vocal minority of American women reflects, above all, the sexist consequences of urban life-styles in American society.

If these symptoms of instability are more grave in American cities than in the cities of social democratic Western Europe, this merely supports the correlation between the severity of capitalist contradictions and the degree of capitalism. The social democratic, mixed economy masks the symptoms by administering palliative medicines, but fails to effect a cure. The underlying exploitation of the many by the few, owing to the concentration of economic power, remains.

Engels's views on urban problems

Marxist analyses of the city are particularly attractive when dealing with situations where they offer a perception quite different from and more accurate than those suggested by traditional interpretations. A very good illustration of this point is urban renewal, and especially the United States urban-renewal programme since the late 1940s. Although this has been criticized by many contemporary observers such as Anderson (1964), the basic effects of the programme had already been anticipated by Engels over a century ago in his analysis of what he called "Hausmannism": "the most scandalous alleys and lanes disappear to the accompaniment of lavish self-glorification by the bourgeoisie on account of this tremendous success, but—they appear again at once somewhere else, and often in the immediate neighbourhood ... The same economic necessity which produced them in the first place produces them in the next place also" (Engels, 1970 edition, pages 69-71). This statement and the more extensive discussions by Engels would not be unfamiliar to those acquainted with the typical urban-renewal project in the United States— the replacement of inner-city, low-income housing by large-scale commercial buildings or upper-income, high-rise apartments, and the consequent relocation of the poor which creates slums elsewhere in the city.

Almost as convincing is the analogy between Engels's description of Manchester in the 1840s and more recent and contemporary cities in developed countries, but especially in the United States: "Manchester contains, at its heart, a rather extended commercial district, perhaps half a mile long and about as broad ... Nearly the whole district is abandoned by dwellers, and is lonely and deserted at night ... With the exception of this commercial district, all Manchester proper, all Salford and Hulme ... are all unmixed working people's quarters, stretching like a girdle, averaging a mile and a half in breadth, around the commercial district. Outside, beyond this girdle, lives the upper and middle bourgeoisie in remoter villas with gardens ... in free, wholesome country air, in fine comfortable homes, passed every half or quarter hour by omnibuses going into the city. And the finest part of the arrangement is this, that the members of the money aristocracy can take the shortest road through the middle of all the labouring districts without ever seeing that they are in the midst of the grimy misery that lurks to the right and left. ... this hypocritical plan is more or less common to all great cities" (Engels, 1962 edition, pages 46–47). The realization that residential segregation (approximated by two annular rings around the CBD) also implies class segregation goes far in explaining central-city–suburban antagonisms, conflicts that are glossed over in the anonymous competitive model of concentric land use.

Rather less convincing is Engels's opposition to home ownership by the poor in cities. This partly reflects the view that home ownership is a symbol of private ownership (though not of the means of production) and represents a 'bourgeoisification' of the working class. But the main objection is a pragmatic one—that owner-occupation restricts mobility and hence weakens the ability of workers to resist the capitalists: "For our workers in the big cities freedom of movement is the prime condition of existence, and landownership can only be a fetter to them. Give them their own houses, chain them once again to the soil and you break their power of resistance to the wage cutting of the factory owners" (Engels, 1970 edition, page 45). Consequently, "with the present development of large-scale industry and towns this proposal (to replace rent with a mortgage instalment) is as absurd as it is reactionary, and ... the reintroduction of the individual ownership of his dwelling by each individual would be a step backward" (*ibid*, page 93). The implication is that the housing problem, like other urban problems, cannot be solved in isolation from dealing with capitalism itself: "As long as the capitalist mode of production continues to exist it is folly to hope for any isolated settlement of the housing question or of any other social question affecting the lot of the workers" (*ibid*, page 71).

The drawbacks with this analysis are well known, though not beyond debate. Home ownership has an appeal to households which transcends classifying it with consumerism, though the role of mortgage institutions gives the term 'ownership' a rather loose meaning. The argument that abolition of capitalism is a precondition for solving the housing question is not confirmed empirically, in the sense that there are some social democratic countries where housing standards are acceptable even to the relatively poor. Moreover capitalist–socialist comparisons of housing consumption levels are hardly flattering to the latter. A stronger case, though ideologically unacceptable because it cuts across the owner–rental issue, can be made for the inequitable distribution of the quality of housing stock in capitalist cities. To the extent that this distribution of housing stock is an inevitable consequence of the distribution of income, it is arguable that the latter could be radically changed only by transforming the economic system.

Town versus country

According to Marx and Engels, one of the main problems associated with urbanization under capitalism is the antithesis between town and country[31]. This reflects their analysis of the nineteenth century world as they saw it. In the modern world, this antithesis has been replaced in developed countries by other contradictions such as those between advanced capitalist and underdeveloped countries, and within cities between the central city and the suburbs. However, in many developing countries—especially in Latin America and in South East Asia—the urban-rural problem is more serious than ever. According to dependency theory (*dependencia*), the urban hierarchy exists as a channel for extracting surplus value from a rural and resource hinterland for shipment to the primate city and other major metropolises, and from them to the international centres of monopoly and financial capitalism and their sibling rivals, the multinational corporations. This is the familiar extension of Marxism in the light of subsequent historical development—implications of the theory of colonialism and imperialism as explained in the writings of Hobson, Lenin, Frank, and others.

Harvey (1973) argues that the non-Marxist and Marxist views about urban-rural relations can be illuminated, if a little simplistically, by Hoselitz's (1954-1955) distinction between 'generative' and 'parasitic' cities. According to the generative concept, the city is beneficial to the countryside because it is the catalyst for economic development and the centre of innovations. Even if urban growth requires the creation and extraction of a food surplus from rural areas, so that cities derive their subsistence and much of their wealth from the surrounding countryside (that is, are sustained by their hinterlands), the relationship is mutually advantageous because of the gains from trade resulting from specialization and the division of labour. The Marxists argue, however, that the extraction of a surplus—not only to feed the city but also to provide the resources for expanding production (investment)—involves exploitation in the form of primitive accumulation. In the parasitic city even the growth function of the ruling class is not carried out, because the surplus is extracted but then used for purposes of conspicuous consumption. The parasitic city becomes the refuge of the landowners, the rentiers, and the bureaucracy who live off the surplus extracted from the countryside.

There is a contradiction even in socialist thinking between the need for a surplus in order to develop society, and the mechanisms by which this surplus is extracted. The solution is that rent, interest, and profit are replaced by socially necessary labour that produces use values rather than

---

[31] This section is included because it was the aspect of spatial economics stressed by Marx and Engels. Also, it is relevant to today's debates on national-settlement strategies. However, it should be pointed out that modern Marxist urban theorists do not take it very seriously, since they are more concerned with the class conflict and and distribution of income *within* cities.

exchange values; that is, under socialism the surplus is created out of unalienated labour. Thus the surplus does not accumulate via class exploitation but by voluntary donation of a quantity of each worker's surplus labour for the social good. Less easy to deal with is Marx's and Engels's assertion that the antithesis between town and country would disappear under socialism. Since agglomeration and other scale economies are an important source of growth, there are considerable advantages in big cities. To let them run down involves social costs. In countries such as Tanzania, which have a low level of urbanization, it may be feasible to promote development and urbanization in a manner which permits a better balance between urban and rural life. In China, and to a lesser extent in Cuba, conscious and strong measures of an economic, social, and political nature are taken to control the growth of the big cities and their bureaucratic dominance, and much of the stress in the development effort is placed on the rural sector. It would be very difficult to unlock the metropolitan stranglehold in countries such as those in Latin America where cities are more developed and have already acquired strong cosmopolitan and capitalist traits.

A more modest and less debatable interpretation is that, instead of trying to abolish the large cities—as often suggested by Engels—the productivity of urban agglomerations should be used for the benefit of the country as a whole by distributing it in the form of services to the population at large, especially in the countryside. Thus national systems of regional health, education, and social welfare programmes become the socialist equivalent of the private consumerism stimulated by monopoly capitalism. The drawback with this argument is that it is a plausible hypothesis that these public redistributive strategies are not beyond the capacity of a social democratic system in which many of the means of production remain in the hands of a heavily taxed private sector.

Marx argued that "the foundation of every division of labour that is well developed, and brought about by the exchange of commodities, is the separation between town and country. It may be said that the whole economic history of society is summed up in the movement of this antithesis" (Marx, 1967 edition, volume III). It was necessary to abolish this in order to save both the urban masses and the rural population. "The present poisoning of the air, water and land can be put an end to only by the fusion of town and country; and only such fusion will change the situation of the masses now languishing in the towns ... Only as uniform a distribution as possible of the population over the whole country, only an intimate connection between industrial and agricultural production together with the extension of the means of communication made necessary thereby—granted the abolition of the capitalist mode of production—will be able to deliver the rural population from the isolation and stupor in which it has vegetated almost unchanged for thousands of years" (Engels, 1970 edition, page 89). Except as a natural consequence of the abolition of capitalism, the writings of Marx and Engels are vague as to how these changes will be effected.

The Chinese and Cuban approaches veer too heavily in favour of rural development to attain a satisfactory balance between town and country. Some suggestions of Soviet architects and planners (Gutnov et al., 1970) for self-contained semirural settlements are not much different from discarded notions about the British New Towns. Engels's prescription of uniform population distribution is infeasible, uneconomic, and socially undesirable. Growth-pole strategies are not inconsistent with socialism, especially if they are concentrated on small towns in small regions (Pióro, 1972; Gohman and Karpov, 1972). There are few insights in Marxist analysis on the formulation of spatial strategies for primate-city size distributions in the third world. Devising complementary policies for dealing with the poor in the big cities and attempting to slow down rural–urban migration by rural development are obvious approaches, but even if socialist revolution were a necessary condition it is certainly not a sufficient condition. Finally, the Marxist view is that the towns exploit the countryside. In developed countries, it is at least as plausible that concentration of ownership and heavy capitalization of agriculture (with relatively few landless labourers) lead to the exploitation of urban consumers by the farm sector, buttressed by discriminatory price-support schemes operated by governments.

Concluding comments

It is difficult to offer even the pretence of a detached assessment of Marxist views of the city and of urban problems, because so much depends upon value judgments. Moreover there is a marked contrast between the diagnosis of urban problems—where the Marxists are frequently acute and penetrating—and prescription, where the Marxist position is very extreme because it offers only an all-or-nothing solution. To argue that measures to tackle urban problems directly are worthless, because they are, at best, merely palliatives and do not get to the root of the problem, is too defeatist in its effects if this line is adopted in a stable, but nonsocialist political system. In countries where the radical perspective receives only fringe support, the impact of the radical position is wasted because its supporters spend too much time in protesting and too little in reconstructing. This is a pity because Marxist analysis of urban ills very often reveals—especially on distributional issues—insights which are not to be found in more traditional diagnoses.

On the other hand a limitation of Marxist urban theory is that it is rather thin, incomplete, and anachronistic. Although the urban system was one of the arenas where the struggle against capitalism was fought, it was not a critical arena for Marx. Moreover the major impacts of

capitalism on urban development have occurred since Marx died. In their
reverence to the master, latter-day Marxists have been slow to extend his
analysis to make it more relevant to the changing environment. There are
a few exceptions to this generalization, for instance, the theory of
monopoly-finance capitalism and of colonialism, but they do not include
the urban economy. Few Marxist writers in English have given much
attention to analysis of the city. David Harvey is a rare exception, though
his analysis is very much an elaboration of Marx rather than an advance
upon it. Lefebvre has perhaps been more innovative in his argument that
the secondary circuit is supplanting the primary circuit, but, leaving aside
the philosophical abstractions, the analysis becomes a Marxist analogue at
the urban level of the changing composition of economic activity away from
manufacturing and towards service industries. The other strand of analysis
in post-Marx urban theory is to argue that the decentralized metropolis,
the domination of the automobile, and the individualistic suburban
life-style are monster creatures of monopoly capitalism. This overstates
the case. It is clear that this type of urban spatial structure is inefficient
in a narrow, economic sense. It is not so clear that it conflicts with the
satisfaction of household and consumer preferences, even after allowing
for the role of advertising and the manipulation of individual tastes by a
conspiracy of industry promotions and societal pressures.

The importance of Marxist analyses of the city, therefore, lies not so
much in the specifics of the argument or in the light it sheds on socialist
ideology but in its general approach and its difference in perspective. The
stress on historical development, on the interrelations between urban
problems and society at large, and on the dark side of the effects of
competitive processes within cities, all give the Marxist diagnoses a
freshness and insight missing from the more formal, and more arid,
neoclassical models. Even when the latter touch upon policy problems,
the typical issue is whether or not the introduction of price adjustments
(for example, taxes and subsidies) would convert an equilibrium allocation
into an optimum. The political feasibility of user charges, congestion
tolls, and other pricing strategies in cities is open to doubt, even in
market-oriented economies such as the United States. Even though
Marxist prescriptions may also be criticized for lack of political reality, at
least in many countries at the present time, the Marxist diagnoses reveal
the adverse distributional side effects of pricing strategies and competitive
solutions. If the Marxist looking glass is distorted, so is that of the
neoclassicist.

2 The unheavenly city

Among other political economy perspectives on the city, none has received
as much attention recently as that of Banfield (1974). At first sight
diametrically opposed to Marx, his analysis leads to similar conclusions
about the effectiveness of liberal reform programmes, though for rather

different reasons. His position is easily summarized: in general, urban problems are not serious, and the few that are serious we can do nothing about. If we tried to do anything about them, we would make matters worse. The arguments have many nuances, but they boil down to these simple statements. The frame of reference is American cities, but Banfield would argue that they also apply, perhaps with a little less force, to cities in other countries.

Many urban problems, such as congestion, are not so much problems as inevitable—and bearable—characteristics of urban life. Others reflect the attitudes of the white middle class, and affect their comfort and convenience rather than being critical to their welfare. Several of the so-called urban problems are more characteristic of rural areas and small towns than big cities (a notable exception is crime). Some urban problems could be solved quite easily, if the measures were acceptable. The price of solving them is political, and unwillingness to pay this price must imply refusal to accept them as critical. For example, much of the so-called revenue crisis could be remedied by charging nonresidents for the services they receive and/or by changing city boundaries. These reforms are politically unacceptable. Thus the urban fiscal crisis "reflects the fact that people hate to pay taxes and that they think that by crying poverty they can shift some of the bill to someone else" (Banfield, 1974, page 8).

Banfield admits that some urban problems are quite serious—crime, poverty, ignorance, and racial injustice. But standards have been rising steadily, and in most cases there have been substantial improvements. Perhaps the trouble is that expectations have been rising faster than the rate of improvement:

"our urban problems are like the mechanical rabbit at the racetrack, which is set to keep just ahead of the dogs no matter how fast they may run. Our performance is better and better, but because we set our standards and expectations to keep ahead of performance, the problems are never any nearer to solution. Indeed, if standards and expectations rise *faster* than performance, the problems may get (relatively) worse as they get (absolutely) better" (*ibid*, pages 23-24).

This creates frustration and resentment, especially if policies to accelerate the improvement are (as argued below) doomed to failure.

The serious problems of the city arise, in Banfield's view, from one source—the existence of a class of "people who live for the present and for whom the present is often empty" (*ibid*, page 72)—the subculture of the poor for whom the slum is a style of life. Or, more explicitly and more frankly, "the lower-class forms of all problems are at bottom a single problem: the existence of an outlook and style of life which is radically present-oriented and which therefore attaches no value to work, sacrifice, self-improvement, or service to family, friends, or community" (*ibid*, page 235). Moreover, this outlook is not the result of society's neglect, it is not due to inadequate job opportunities, poor schooling or discrimination, but it is almost inbred: "the lower-class person has been permanently damaged by having been assimilated in infancy and early childhood into a pathological culture" (*ibid*, page 236). If the 'do-good' interventionism, of which Banfield is so critical, smacks of paternalism, this smacks of contempt.

If most urban ills arise from this source, nothing can be done about them, at least, directly and in the short run. The best chances of improvement lie in the natural development of spontaneous forces— economic growth, demographic change (especially the falling birthrate), and upward social mobility. Massive government programmes to aid the cities will be ineffective, and though most of them "will lead to nothing worse than some additional waste and delay", some are counterproductive. Two important examples in recent American experience are urban transportation policies and housing policies (particularly urban renewal and subsidies to home owners), which have stimulated out-migration from central cities. The range of feasible and acceptable policy options is very narrow, the measures adopted are perverse (mainly because of the nature of American political institutions and the influence of the upper-class cultural ideal of 'service' and responsibility to the community), and the introduction of these useless measures will create a "bureaucratic juggernaut" which will perpetuate itself. Physical planning prescriptions and institutional reforms cannot work because they do not attack the real problem: "even if we could afford to throw the existing cities away and build new ones from scratch, matters would not be essentially different, for the people who move into the new cities would take the same old problems with them" (*ibid*, page 280).

To the extent that Banfield offers policy suggestions, they are far from the typical, liberal urban strategies. He suggests equality of access, including access to housing and jobs; this is common ground with the liberals. The others are not. They include: reducing years of schooling; institutionalization of the "incompetent poor"; aggressive birth-control programmes; training of the children of problem families in nurseries to adjust them to "normal culture"; more police and tougher punishments; and prohibition of television coverage of riots. This odd mixture of proposals has totalitarian undertones, and it is hardly surprising that Banfield's analysis has been the object of strong attacks.

Many of the objections to Banfield attack his cynicism. In its place, however, they substitute pious optimism. The following quotation is not untypical:

"To attain justice with fairness, what is needed is a broad moral consensus, founded in a heightened perception of human interdependence and an understanding that a just and free society are the stakes of a game which must be played out within a viable time frame" (Bateman and Hochman, 1971, page 352).

To observers such as these, the fault is not that of the lower classes themselves, but of the political majority for not adopting the essential reforms in social and economic institutions. However, as Banfield argued, the reason why the political majority have neglected to do this is because they wanted to; the costs would be too high, and the problems are not critical enough. To exhort society to think of its long-term interests (even assuming that drastic action to deal with urban problems is consistent with those interests) is very different from making society act. Banfield may have been right when he argued that the policies that would work are unacceptable, and those that are acceptable would not work.

Despite the insights and frankness of the Banfield analysis, it would be unnecessarily defeatist to accept his conclusions. There is a middle ground between waiting (possibly indefinitely) for problems to solve themselves and acting in vain to create utopia now. Experience in other countries suggests that urban policies and planning may be partially successful. The American political structure tends to dilute the effectiveness of action, from the legislative to the implementation stage. But the urban-renewal policy stands for a polar extreme rather than the norm. Unless the struggle for improvement is continued, Banfield's analysis, like that of Marx when applied to the United States, leads not only to inaction but to cynicism and despair. His stress on the innate character defects of what he calls the "lower class" is no less socially divisive than Marx's attack on bourgeois capitalists. It is paradoxical that the more prominent writers on urban political economy—presumably, a field stressing the interactions between political and economic change—should so frequently come out in favour of no urban policies, because they believe either that no measures would work or that the only workable strategy is a total revolution. Even the faith of NUE and other neoclassical theorists in the effectiveness of the price mechanism and market incentives is more constructive than that.

14

Conclusion: are NUE models operational?

There would be no value in a theory which closely represented reality in detail. The essence of theory is simplicity, and economists tend to judge the quality of a theory by its simplicity and elegance. If theory is studied for its own sake, this criterion makes sense. An economic model—like a master-game of chess—may be admired for its beauty. But if theories are also to be judged by their operational potential and by their value to planners and policymakers, the criteria have to be broader. In particular, the overlap between simplicity and relevance may be very small. NUE models are undoubtedly a radical simplification of that complex phenomenon—the modern city. The assumptions common to most of the models are rather drastic: a monocentric city; rigid segregation of land uses with production in the CBD and residences in the surrounding rings; the ubiquity of transportation routes; the absence of locational interdependence; the continuity and smoothness of rent and density gradients; the underlying reliance on competitive forces, marginal adjustments and, at best, a passive role for a planning authority.

These simplifications would have been acceptable if they had been a prerequisite for the formulation of a testable model, and if the tests carried out had yielded sound predictions. There have been no such tests. The closest approximation has been simple numerical solutions which have, in the opinion of the designers of the model, generated plausible findings compatible with the predictions of the older, partial, and less mathematical theories and with available empirical evidence. However, these predictions tend to be too general; for example, a negative rent gradient, or the poor living closer in than the rich. Moreover, as pointed out earlier, some of these general predictions are not as universally applicable as is sometimes thought. To this extent the capacity to generate the standard predictions is a weakness, and an indication of the theory's restrictiveness or the theorists' blinkered outlook, rather than testimony to its soundness.

A related problem is that if a theoretical framework is being judged for policy relevance, assumptions that oversimplify—a plus point for neat and elegant theory—may be unhelpful. First, policy is concerned, *inter alia*, with identifying strategic levers for bringing about change, and it is necessary to quantify the impacts that follow from exerting specific degrees of pressure to these levers. The qualitative results so highly prized by pure theorists are insufficiently precise for the policymaker. Second, urban policy has always to keep locational interdependencies at the forefront. The more important of these are assumed away in NUE. Third, policy is applied in a particular institutional context. NUE either ignores this or makes implicit institutional assumptions that frequently conflict with reality.

Conclusion: are NUE models operational?

An interesting analogy has been drawn between NUE and macroeconomics. This is that the abstract models of NUE may eventually be justified because they will lead to practical, useful policy models, just as basic macroeconomic theory led to the development of national short-run forecasting models. "The relation seems closely analogous to that in macroeconomics, between the large econometric models on the one hand and the tradition of small-scale macroeconomic theory—whether mathematical, graphical or verbal—on the other" (Solow, 1973b, page 267).

This argument relies more on fond hope than on reasoned logic. First, there are big differences between macroeconomic and urban (especially NUE) models, which make the transition to applications much more problematic in the latter case. The macroeconomic models (1) are linear; (2) use more easily measurable variables; (3) include simple, key control variables (for example the level of government expenditures and the tax rate) that have no direct parallel in NUE; and (4) are nonspatial, and hence are much easier both to model and to calibrate.

Second, there is another theoretical subfield in macroeconomics which is even less encouraging to the 'progress-to-applications' thesis. This is, of course, aggregate growth theory. Is NUE closer to the simple Keynesian macroeconomic model or to the aggregate growth models? The parallels between NUE and aggregate growth theory are very close: a standard model where slight changes in the assumptions yield marginally differentiated results; the same mathematical tools (calculus of variations and optimal control theory); a similar degree of abstraction with models that are notable, above all, for the stark—though attractive—simplicity of their assumptions; and the two games even share some of the same players. Aggregate growth theory has not led to applications, and it has not been helpful to policymakers. Its main output has been displays of 'neoclassical pyrotechnics', an observed feature of the early NUE models (Mills and MacKinnon, 1973).

Though we cannot expect a general theoretical framework to be directly useful for the analysis of detailed planning problems, it would be helpful if the theory could be expressed in terms (level of disaggregation, treatment of time dimension, etc) within which these problems could, at least, be discussed. Many major planning problems fall into two categories: (1) predictions of overall urban change by using a forecasting model; (2) studies of changes in the spatial structure at the intraurban level (for example, the dynamics of land-use competition between residences and nonresidences; neighbourhood upgrading and blight; the design of transportation systems, capacities, and networks; the growth and planning of secondary subcentres within metropolitan areas).

NUE has few insights, however broad, to offer on these questions, primarily because the models do not include time and they trivialize space (one dimension rather than two or, better still, three). Despite the limitations of simple planning models, such as the Lowry (1964) model, they have demonstrated their usefulness to planners by providing approximate forecasts of the spatial allocative changes consequent upon given increments of overall urban growth, even if they have failed to say very much about the nature of the adjustment paths. The closest that NUE models can come to forecasting is by exploring parameter changes in a sensitivity analysis by using numerical methods.

The value of NUE for intraurban analysis is severely restricted by its assumption of a one-dimensional linear ray, hence its inability to deal with intraurban zones. In fact zones would not have much of a role in NUE models because of its rigid land-use segregation. It is difficult to justify its claimed virtue as a *competitive* model of land use because it rules out, by assumption, the more important aspects of this competition—namely, the competition between residential and commercial use. Competition for land in the NUE model is between housing and roads. But this is a distorted competitive market because road development is a function of the public sector, and land is frequently acquired compulsorily at purchase prices that are not 'true' market prices.

In my view a satisfactory competitive model of land use would require: (1) two-dimensional zones that would permit analysis of neighbourhood change and the growth of subcentres via nonresidential takeovers or expansion of noncentral sites; (2) different types of nonresidential land use with varying requirements of site and accessibility (large commercial and office establishments mainly in the CBD; large-scale retail establishments, either in the CBD or at suburban agglomerations; manufacturing, competing with housing and agriculture on the urban fringe; small retail establishments interspersed with residences at varying distances from the CBD); (3) due allowance for the constraints on the operation of market forces: (a) occupation of most sites, sluggishness of change due to long leases and locational inertia, etc; (b) the degree of control exerted over the urban land market by the planning authority (zoning, lot size standards, building codes, and planning permits); and (c) imperfections in competition for other economic and social reasons, such as residential-segregation theory, dual housing and labour markets for blacks and whites, differences in the environmental quality in neighbourhoods which affect property values irregularly with distance, heterogeneity of housing (ignored because housing demand is usually treated as the demand for land), and locational interdependence. The argument, of course, is not that all these considerations should be taken into account in detail, for that would be to expect theory to replicate reality, but rather that urban models should focus on the key elements of land-use competition and that they should recognize the existence of constraints on competition (inertia constraints, adjustment lags, etc). These features can probably be handled much more easily with, say, a system-simulation model than with a neoclassical competitive-equilibrium model.

A major weakness of the standard NUE model is its assumption of monocentricity and the corollary of a single CBD workplace; "the assumption of single employment center has been maintained by the urban economists in the face of increasing decentralization, and a decline of the role of the CBD as the single focus of productive activity. It has survived despite its increasing irrelevance because it produces simplifications in the analysis and permits important results to be derived which would be difficult to obtain otherwise ... this assumption leads to unrealistic conclusions about the spatial structure of cities. Ultimately, it must be rejected in order to remove inconsistencies from the theory" (Angel and Hyman, 1972a, page 105). The few empirical studies that have been carried out tend to reinforce this conclusion. For instance, Angel and Hyman themselves found that the transport expenditures incurred by worker-households at different residential locations were very different from those to be expected if all jobs were centralized, with transport costs incurred by households close to the CBD much higher than those in the monocentric model, whereas households living at distances far from the city centre incurred, on average, much lower transport expenditures than predicted by the monocentric model. These findings imply that many workplaces are decentralized. The 'exclusive-zoning hypothesis', which states that workplaces are confined to a central core surrounded by a residential ring, was found deficient, both in nineteenth century Chicago (Fales and Moses, 1972) and in a sample of modern American cities (Capozza, 1976). Empirical estimates of employment density gradients (Mills, 1972a; Kemper and Schmenner, 1974) confirm the existence of decentralized workplaces and the fact of increasing decentralization over time. These empirical findings have yet to have much of an impact on the reformulation of NUE theories.

There are two somewhat different approaches to relaxing the assumption of the monocentric city. One is to develop a multicentric locational framework, the other is to construct a model of a more general decentralized city where workplaces are located outside the CBD though not necessarily clustering to form subcentres. The complications introduced with multicentricity are severe, particularly from the point of view of mathematical tractability. The single workplace is a key assumption for compressing two dimensions into one, unless jobs are equally distributed radially (an implausible case). If multiple centres are permitted, an early extension is to permit specialization of function among an intraurban hierarchy of centres, and this means abandoning the analytical convenience of the assumption of a composite consumption good. With multiple goods and specialization among centres, a satisfactory model would need to accommodate intrametropolitan freight shipments as well as coping with more complex commuting patterns. It consequently becomes much more difficult to obtain determinate solutions, and rent and density surfaces may no longer be smooth and differentiable.

Papageorgiou and Casetti (1971) and Papageorgiou (1974) have carried out some preliminary work on residential spatial structure in a multicentric system, in the latter study for a continuous income distribution. Hartwick and Hartwick (1974) generalized a Mills-type linear-programming model to deal with multiple centres and intermediate goods. Lave (1974) showed that decentralization becomes efficient when freight and commuting cost and rent savings outweigh agglomeration economies. Interestingly, his model suggests that once the monopoly of the CBD breaks up, several centres rather than merely two develop; this is consistent with what happens in the real world. The model tends to overproduce centres, but this is due to its simplified treatment of the city as a workplace, production locus, and shopping centre rather than as a complex system for economic and social interaction. The failure hitherto to develop a fully satisfactory subcentring model is probably due to our incomplete understanding of the agglomeration process. To explain the decline of the CBD it is important to know why it grew in the first place. Although there has been much diffuse analysis of agglomeration economies, and a few interesting insights into the 'public-goods' aspects of city centres (for example, Artle, 1973), most NUE modellers—with a few exceptions (Dixit, 1973; Livesey, 1976; Alao, 1976)—have treated the CBD as having no intrinsic interest, merely a dumping ground for commuters. Since the growth of secondary centres in a metropolitan area is part of a dynamic process, the deficiencies of static models are quite glaring on this point.

More general decentralized land-use models have not generated much attention. Niedercorn (1971) showed how the nonresidential demand for land may generate a negative exponential rental gradient, so that allowing for mixed land uses need not destroy the basic findings of the standard model. Beckmann (1976) allows jobs to be distributed in the same way as population and demonstrates that introducing a substitute agglomerating factor (the need for social interaction, cf Artle's 'agora model', 1973) still leads to a negative density gradient, though the 'peaking' at the centre is much less pronounced than in the single CBD workplace model. There remains much work to be done in this area. These observations suggest that there is still so much to be achieved by extensions of the theory that one justification for the slow pace of applications thus far is that testing incomplete models would be premature.

One of the chief bases of support for NUE models is that they produce results consistent with data from real cities and with the findings of earlier urban theory. An example worth examining is that of the negative residential rent gradient that is accepted by NUE and most other urban modellers as beyond dispute, apart from minor quibbles as to whether the rent function is exponential or not (Hochman and Pines, 1971).

In fact there are several reasons why the residential-rent function may be positive, or at least much more irregular than is usually implied. First, although the origins of the negative rent function derive from the von Thünen model, it has recently been shown (Artle and Varaiya, 1974) that this does not follow from von Thünen's own assumptions. Rents (profits) may increase with distance if money wages decline with distance faster than transport costs. von Thünen assumed that real wages are constant everywhere, and it is this assumption which opens the door to positively-inclined sections on the rent gradient (money wages fall with distance because food is more expensive close to the city centre).

Second, some empirical research has shown that residential property values (and rents) may increase with distance (Wilkinson, 1972; Ridker and Henning, 1967; Richardson et al., 1974a). In the last-named study, for instance, a quadratic (inverted-U) distance function yielded the best fit. One explanation of the predominance of negative rent gradients is the competition of nonresidential land uses, which boosts the demand for land close to the city centre. This competition is expressly ruled out in NUE models. Another important factor is that NUE in particular, and urban land-rent theory in general, treat urban rent solely as location rent, with increasing distance from the CBD being associated with reductions in accessibility and hence with lower rents.

Thus a third objection to the prediction of a negative rent gradient is that it derives from a very narrow view of the determinants of urban residential rents. There may be other influences on residential property values and rents that tend to make them increase with distance. An important set of influences of this kind are those characteristics associated with the quality of the environment and the neighbourhood. For operational purposes, these may be represented by the surrogate measure of low density (Wabe, 1971; Mirrlees, 1972). The evidence for a negative density gradient is very strong. Theoretically, therefore, a decline in location rent with increasing distance is associated with rising "externality rent" (Richardson, 1977). In other words, a fall in accessibility to the CBD is accompanied by the countervailing advantage of an improvement in the quality of the neighbourhood and the environment. Total rents may decline or increase with distance according to the relative weight of the two effects, which depends on the values of the parameters. It is possible to introduce externality rent and variables relating to environmental quality into NUE models, and there have been a few hints of how this might be done (Papageorgiou, 1976b; Richardson, 1977). However, the fact that such models have not yet been fully analyzed suggests that the theory is still in an underdeveloped state and not yet ripe for empirical testing.

My own prejudice is that urban economics should be primarily a policy-oriented field, and I am not convinced that NUE has much to offer as a guide to policymakers. Indeed, NUE has already attracted some urban economists away from the policy end of the spectrum towards the theory end. That would be all right if the theories developed had important policy variables in them. Apart from policy instruments that deal with a very abstract model of the transportation system, pollution and congestion levies, and a few other pricing strategies there are few signs of this type of theory in NUE. On the contrary the choice of structure of NUE models appears to have been determined by what is mathematically manipulable rather than by what would aid policymakers the most. When NUE models contain policy instruments these are almost always the pricing tools and the user charges that betray their neoclassical origins.

A similar problem is that in order to obtain solutions to the differential equations of NUE models, the theorists often have to assume that particular functions are of a very simple and highly specific form (for example, the logarithmic utility function). The simulation models, on the other hand, allow considerable experimentation with the parameters and the form of functional relationships once the basic model has been set up. Examples include the elasticities of demand for housing, technology (in housing, production, or transportation), and the structure of the transportation system. These specifics determine the outcome when a policy instrument is introduced. The kind of prediction possible with an NUE model, for example, that a congestion toll on road use will result in a smaller city, is too general to be of practical help. The real question is how the spatial structure of the city would change, over what time period, and in what particular ways.

Pessimism about the operational potential of NUE is reinforced by the belief that some of the other mathematical approaches to urban analysis offer much more scope for applications. These alternative approaches are frequently denigrated by economic theorists because they are 'inelegant' and 'empirical'. That may be so, but they have the offsetting advantages that they can be, and in many cases have been, applied, and that they are useful to planners and urban policymakers. Examples include the linear-programming models suggested by Mills (1972b), the Lowry model and its descendants (Goldner, 1971), Wilson's 'comprehensive' models (Wilson, 1974), and Baumol's dynamic cumulative-deterioration model (Baumol, 1963; Oates *et al.*, 1971).

Mills's approach has several advantages: division of the city into zones (square grids) deals with two-dimensional space in a nontrivial way; capital–land substitution is allowed so that the city can grow upwards as well as outwards; production is endogenous; and institutional constraints can be accommodated. These benefits offset the loss of theoretical purity, the need for real data, and the reliance on linear approximations to changes in cost functions.

The simplicity of the Lowry model has not prevented interesting and useful applications. Although the model has been refined and disaggregated since it was first introduced in 1962–1963, its essential characteristics remain: the generation of total employment and its spatial distribution from exogenously-determined basic employment; the generation of residential locations from the work-to-residence trip by using a gravity model; and constraints on these allocations in the form of constraints on the maximum, zonal residential density and minimum threshold sizes for service-employment clusters at the neighbourhood, local, and metropolitan levels.

The Wilson 'comprehensive' approach is cumbersome but effective. The analysis is based on the assumption that urban modelling requires the specification of interrelated submodels that refer to population, economic activity, transportation, and location. Particular emphasis is placed on residence–workplace interdependencies. The models are intended to be heavily spatially disaggregated and to deal with urban change over a single discrete time period. For practical planning purposes Wilson stresses the importance of keeping models relatively simple "to match data availability".

Baumol's cumulative-deterioration models have rarely been applied (an exception is simple econometric testing; Bradford and Kelejian, 1973), yet their emphasis on positive feedback loops suggests a ready use in system-simulation models of the Forrester type. Conceptually the models are attractive because their emphasis on dynamic disequilibrium processes (with a low-level equilibrium trap) contrasts strikingly with the competitive equilibrium of NUE models. Also, they can be used to focus either on individual neighbourhood change or on central-city–suburban relations. Moreover they concentrate attention on policy variables and on the differential effects of various kinds of policies. The rationale of the models is "the dynamic, cumulative nature of the forces which appear to compound the difficulties of the cities from period to period. One finds in a variety of elements of urban life that the process of change involves obvious feedback relationships which reinforce one another and are likely to generate cumulative movements over time" (Oates et al., 1971, page 142). With refinement such approaches have considerable potential as a framework for applied urban-policy models, especially for declining or rapidly expanding cities.

Many of the alternatives to NUE fall within the classification of simulation models. The simulation approach has received a bad press among economists; there are several reasons for this. Economists are traditionally more familiar with, and have a strong preference for, econometric models. One example from the simulation model group—Forrester's *Urban Dynamics* (1969)—has tarnished the reputation of simulation models. Some economists have criticized simulation models because "no one can fully understand what is happening in the bowels of the machine" (Solow, 1973b, page 267). This argument is weak since the models are built up from individual blocks, the structural equations of which have to be very specific. Similarly, the related argument that it is difficult to know what is going on if simultaneous changes are introduced is also overstated, since tests can be controlled by varying parameters sequentially and in combination. A further misplaced argument is that major modifications to simple assumptions should not be made simultaneously but should be introduced one by one, even on an *ad hoc* basis. However, interdependencies in urban economies are so great that fewer mistakes may be made if major problems are handled simultaneously, as they can easily be in a simulation model. The behaviour of NUE models is very sensitive to their assumptions, so that the *ad hoc* handling of one problem may solve one difficulty, but at the same time create another.

It may be arguable that NUE theory and simulation models are complementary: "The building blocks of the big models come from the theory, and from single-equation testing. When they are put together in a big model, anomalies sometimes appear which pose new theoretical questions. Tentative theoretical answers go back into the big models, because pencil-and-paper ones are inadequate for direct application" (Solow, 1973b, page 267). There is some value in this argument, but NUE models may not provide the most appropriate theory. The reason is that they assume narrowly economic rational responses, whereas complex behavioural responses are more relevant and can be simulated relatively easily in a computer model. NUE theories abstract from adjustment lags, constraints on competition, and locational inertia. These make a difference to the predictions and can be handled without trouble within the framework of a simulation model.

On the other hand, the difficulties of simulation models should not be underestimated. The experience with them has hardly been a series of undiluted successes. For example, not too much confidence should be attached to a model simply because it satisfactorily replicates reality in the test-bed city. A test of the model in other cities may quickly reveal misspecifications.

To clarify the standpoint taken here, NUE models are being judged solely on the criterion of whether they will be applied and shown to be relevant to planners and policymakers. Some NUE theorists would argue that this was never their intention. I have no quarrel with this position. Although I remain sceptical of a theory based on assumptions that seem more relevant to an analysis of the nineteenth century rather than the modern city (though even in the former case, the standard model may work poorly, see Fales and Moses, 1972), I respect the right to play this particular game. Moreover, my criticisms of some years ago now seem too harsh (Richardson, 1973a), and the recent progress in NUE has been quite remarkable.

For instance, there have been some attempts to remedy the restrictiveness of static analysis. Although these attempts have not been conspicuously successful, at least the problems of confronting dynamics have been recognized. For example, Anas (1976a) identifies the critical questions of durability (the constraints upon factor substitution) and the influence of future expectations on current decisions relating to housing (and other urban) expenditure, whereas Pines (1976) indicates how the apparently insuperable problem of handling time and space simultaneously in a continuous model may be made a little more tractable by treating one dimension continuously and the other discretely. Muth (1976), and a not dissimilar earlier study by Evans (1975), illustrates a different approach: the development of nonspatial models of urban growth similar in structure to the much more familiar models of aggregate growth. These models are then used to draw inferences about the dynamics of urban spatial structure. Building upon work by Bussière (1972), Mogridge (1974) has shown how a simple concept such as the density gradient may be made dynamic by allowing its main parameters to be functionally related to population and income growth. All of this work is very preliminary, but already it suggests that dynamic models will yield different qualitative results from the standard static models. However, it hardly needs stressing that the extension of the theory into dynamics—desirable and necessary as it is—is likely to postpone the prospect of applications even further into the future.

Thus, I see little prospect of NUE being applied in the sense of offering direct insights into the solution of urban problems or of being much help to planners. The analogy with growth theory comes to mind again. This prestigious field of economic theory did not help to explain why some countries grew faster than others, what were the underlying determinants of growth, or how policymakers could raise a country's rate of growth. This is not a pejorative statement. I am not implying that it was meant to do these things, merely stating the fact that it never did. Indeed, I would argue that it could not. Just as with NUE, the straitjacket of the model's assumptions—needed to obtain determinate solutions—rules out the possibility of applications. I expect NUE to flourish and to make interesting new theoretical discoveries. I do not expect it to have increasing contact with operational models, and I doubt whether NUE will disseminate among planners and urban policymakers. In other words, I expect NUE and applied urban analysis to advance in parallel rather than to converge. Nevertheless, progress in urban economics will be much stronger if the theorists and policymakers maintain contact and try to learn from each other.

References

* Reference not specifically quoted in the text but of particular relevance to this book.
* Alao N, 1974 "An approach to intraurban location theory" *Economic Geography* **50** 59-69

   Alao N, 1976 "On some determinants of the optimum geography of an urban place" in *Essays in Mathematical Land Use Theory* Ed. G J Papageorgiou (Lexington Books, D C Heath, Lexington, Mass.) pp 199-214

   Alonso W, 1960 "A theory of the urban land market" *Papers and Proceedings of the Regional Science Association* **6** 149-157

   Alonso W, 1964 *Location and Land Use* (Harvard University Press, Cambridge, Mass.)
* Amson J C, 1972 "Equilibrium models of cities: 1. An axiomatic theory" *Environment and Planning* **4** 429-444
* Amson J C, 1973 "Equilibrium models of cities: 2. Single-species cities" *Environment and Planning* **5** 295-338

   Amson J C, 1974 "Equilibrium and catastrophic modes of urban growth" in *London Papers in Regional Science 4. Space-Time Concepts in Urban and Regional Models* Ed. E L Cripps (Pion, London) pp 108-128

   Amson J C, 1976 "A regional plasma theory of land use" in *Essays in Mathematical Land Use Theory* Ed. G J Papageorgiou (Lexington Books, D C Heath, Lexington, Mass.) pp 99-116
* Anas A, 1973 "A dynamic disequilibrium model of residential location" *Environment and Planning* **5** 633-647

   Anas A, 1976a "Dynamics of residential growth" *Journal of Urban Economics* (forthcoming)

   Anas A, 1976b "Short-run dynamics in the spatial housing market" in *Essays in Mathematical Land Use Theory* Ed. G J Papageorgiou (Lexington Books, D C Heath, Lexington, Mass.) pp 261-275

   Anas A, Dendrinos D S, 1976 "The new urban economics: a brief survey" in *Essays in Mathematical Land Use Theory* Ed. G J Papageorgiou (Lexington Books, D C Heath, Lexington, Mass.) pp 23-51

   Anderson M, 1964 *The Federal Bulldozer: A Critical Analysis of Urban Renewal* (MIT Press, Cambridge, Mass.) pp 1949-1962

   Anderson R, Crocker T, 1971 "Air pollution and residential property values" *Urban Studies* **8** 171-180

   Angel S, Hyman G M, 1972a "Urban transport expenditures" *Papers and Proceedings of the Regional Science Association* **29** 105-123
* Angel S, Hyman G M, 1972b "Urban spatial interaction" *Environment and Planning* **4** 99-118

   Arrow K J, 1962 "The economic implications of learning by doing" *Review of Economic Studies* **29** 155-173

   Artle R, 1973 "Cities as public goods" Memorandum number ERL-M417, Electronics Research Laboratory, University of California, Berkeley

   Artle R, Varaiya P, 1974 "On the existence of positive rent gradients in Thünen models" Memorandum number ERL-M459, Electronics Research Laboratory, University of California, Berkeley

   Artle R, Varaiya P, 1975 "Economic theories and empirical models of location choice and land use: a survey" *Proceedings of the Institute of Electrical and Electronics Engineers* **63** 421-430
* Averaus C P, Lee D B, 1973 "Land allocation and transportation pricing in a mixed urban economy" *Journal of Regional Science* **13** 173-185

   Bailey M J, 1959 "Note on the economics of residential zoning and urban renewal" *Land Economics* **35** 288-292

   Banfield E C, 1974 *The Unheavenly City Revisited* (Little, Brown, Boston)

Barr J L, 1972 "City size, land rent and the supply of public goods" *Regional and Urban Economics* **2** 67-103
• Barr J L, 1973 "Tiebout models of community structure" *Papers of the Regional Science Association* **30** 113-139
Bateman W, Hochman H M, 1971 "Social problems and the urban crisis: can public policy make a difference?" *American Economic Review* **61** 346-353
Batty M, 1971 "Design and construction of a sub-regional land use model" *Socio-Economic Planning Sciences* **5** 97-124
Batty M, 1972 "Recent developments in land-use modelling: a review of British research" *Urban Studies* **9** 151-177
• Baumol W J, 1963 "Interactions of public and private decisions" in *Public Expenditure Decisions in the Urban Community* Ed. H G Schaller (Johns Hopkins University Press, Baltimore, for Resources for the Future Inc.) pp 1-18
Becker G S, 1965 "The theory of allocation of time" *Economic Journal* **75** 493-517
Beckmann M J, 1957 "On the distribution of rent and residential density in cities" Interdepartmental Seminar on Mathematical Applications in the Social Sciences, Yale University, New Haven (mimeo)
Beckmann M J, 1958 "City hierarchies and the distribution of city sizes" *Economic Development and Cultural Change* **6** 243-248
Beckmann M J, 1968 *Location Theory* (Random House, New York)
Beckmann M J, 1969 "On the distribution of urban rent and residential density" *Journal of Economic Theory* **1** 60-67
Beckmann M J, 1972 "von Thünen's model revisited: a neoclassical land use model" *Swedish Journal of Economics* **74** 1-7
Beckmann M J, 1973 "Equilibrium models of residential location" *Regional and Urban Economics* **3** 361-368
• Beckmann M J, 1974 "Spatial equilibrium in the housing market" *Journal of Urban Economics* **1** 99-108
• Beckmann M J, 1970 "Equilibrium vs. optimum: spacing of firms and patterns of market areas" (mimeo)
Beckmann M J, 1976 "Spatial equilibrium in the dispersed city" in *Essays in Mathematical Land Use Theory* Ed. G J Papageorgiou (Lexington Books, D C Heath, Lexington, Mass.) pp 117-125
Ben-Shahar H, Mazor A, Pines D, 1969 "Town planning and welfare maximization: a methodological approach" *Regional Studies* **3** 105-113
• Blumenfeld D E, 1972 *Effects of Road System Designs on Congestion and Journey Times in Cities* PhD thesis, University College London, London
Blumenfeld H, 1954 "The tidal wave of metropolitan expansion" *Journal of the American Institute of Planners* **20** 3-14
• Bollobas B, Stern N, 1972 "The optimal structure of market areas" *Journal of Economic Theory* **4** 174-179
Borukhov E, 1973 "City size and transportation costs" *Journal of Political Economy* **81** 1205-1215
• Borukhov E, 1975 "The effects of public provision of roads on the structure and size of cities" *Environment and Planning A* **7** 349-355
Böventer E G von, 1970 "Optimal spatial structure and regional development" *Kyklos* **23** 903-924
Bradford D F, 1972 "Comment on Engle et al." *American Economic Review* Papers **62** 99-102
Bradford D F, Kelejian H, 1973 "An econometric model of the flight to the suburbs" *Journal of Political Economy* **81** 566-589
• Brown H J, Ginn J R, Ingram G K, Kain J F, 1972 *Empirical Models of Urban Land Use* (Columbia University Press, New York, for the National Bureau of Economic Research)

References

- Buchanan J M, Goetz C J, 1972 "Efficiency limits of fiscal mobility: an assessment of the Tiebout model" *Journal of Public Economics* **1** 25-43
- Burgess E W, 1925 "The growth of a city" in *The City* R E Park, E W Burgess, R D MacKenzie (University of Chicago Press, Chicago) pp 47-62
- Bussière R, 1972 *Modèle Urbain de Localisation Résidentielle* (Centre de Recherche d'Urbanisme, Paris)
- Bussière R, Snickars F, 1970 "Derivation of the negative exponential model by an entropy-maximizing method" *Environment and Planning* **2** 295-301
- Capozza D R, 1973 "Subways and land use" *Environment and Planning* **5** 555-576
- Capozza D R, 1976 "Employment-population ratios in urban areas: a model of the urban land, labour and goods markets" in *Essays in Mathematical Land Use Theory* Ed. G J Papageorgiou (Lexington Books, D C Heath, Lexington, Mass.) pp 127-143
- Casetti E, 1969 "Alternative urban population models: an analytical comparison of their validity range" in *London Papers in Regional Science 1. Studies in Regional Science* Ed. A J Scott (Pion, London) pp 105-113
- Casetti E, 1970a "Urban population density patterns: an alternative explanation" *Canadian Geographer* **11** 96-100
- Casetti E, 1970b "Spatial equilibrium distribution of 'rich' and 'poor' households in an idealized urban setting" DP 13, Department of Geography, Ohio State University, Columbus
- Casetti E, 1971 "Equilibrium values and population densities in an ideal setting" *Economic Geography* **47** 16-20
- Casetti E, 1973 "Urban land values: equilibrium vs. optimum" *Economic Geography* **49** 357-365
- Casetti E, 1974 "Spatial equilibrium in an ideal urban setting with continuously distributed incomes" in *London Papers in Regional Science 4. Space-Time Concepts in Urban and Regional Models* Ed. E L Cripps (Pion, London) pp 129-140
- Casetti E, Papageorgiou G J, 1971 "A spatial equilibrium model of urban structure" *Canadian Geographer* **15** 30-37
- Castells M, 1970 "Structures sociales et processus d'urbanisation" *Annales, Économies, Sociétés, Civilisation* **25** 1155-1199
- Castells M, 1973 *La Question Urbaine* (Maspero, Paris)
- Castells M, Godard F, 1974 *Monopolville* (Mouton, Paris)
- Cesario F J, 1975 "A primer on entropy modelling" *Journal of the American Institute of Planners* **41** 40-48
- Clark C, 1951 "Urban population densities" *Journal of the Royal Statistical Society, Series A* **114** 490-496
- Clark C, 1967 *Population Growth and Land Use* (Macmillan, London)
- Cohen B, 1971 "Another theory of residential segregation" *Land Economics* **47** 314-315
- Cordey-Hayes M, 1974 "On the feasibility of simulating the relationship between regional imbalances and city growth" in *London Papers in Regional Science 4. Space-Time Concepts in Urban and Regional Models* Ed. E L Cripps (Pion, London) pp 173-195
- Courant P N, 1976 "On the effect of fiscal zoning on land and housing values" *Journal of Urban Economics* **3** 88-94
- Curry L, 1967 "Central places in the random spatial economy" *Journal of Regional Science* **7** 217-238
- Dacey M F, 1968 "A model for the areal distribution of population in a city with multiple population centers" *Tijdschrift voor Economische en Sociale Geografie* **59** 232-236
- Davies G W, 1974 "The effect of a subway on the spatial distribution of population" Research Report 7404, Department of Economics, University of Western Ontario, London, Ontario

Davis E, Swanson J G, 1972 "On the distribution of city growth rates in a theory of regional economic growth" *Economic Development and Cultural Change* **20** 495-503

Delson J K, 1970 "Correction on the boundary conditions in Beckmann's model of urban rent and residential density" *Journal of Economic Theory* **2** 314-318

Deutsch K W, Isard W, 1961 "A note on the generalized concept of effective distance" *Behavioral Science* **6** 308-311

Devletoglou N E, 1971 *Consumer Behaviour: An Experiment in Analytical Economics* (Harper and Row, London)

Dhrymes P J, 1965 "Some extensions and tests for the CES class of production functions" *Review of Economics and Statistics* **47** 357-366

• Dixit A, 1971 "Route choice and congestion in urban transport" Oxford University, Oxford (mimeo)

Dixit A, 1973 "The optimum factory town" *Bell Journal of Economics and Management Science* **4** 637-651

• Edel M, 1971 "Urban renewal and land use conflicts" *Review of Radical Political Economics* **3** 76-89

Edel M, 1972 "Land values and the costs of urban congestion: measurement and distribution" in *Political Economy of the Environment: Problems of Method* École Pratique des Hautes Études, VIe Section (Mouton, The Hague) pp 61-90

• Edel M, Sclar E, 1974 "Taxes, spending and property values: supply adjustment in a Tiebout-Oates model" *Journal of Political Economy* **82** 941-954

Ellickson B, 1971 "Jurisdictional fragmentation and residential choice" *American Economic Review* Papers **61**, 334-339

Engels F, 1962 edition *The Condition of the Working Class in England in 1844* (Macmillan, London)

Engels F, 1970 edition *The Housing Question* (Progress Publishers, Moscow) originally published in 1872-1873

Engle R F, 1974 "Issues in the specification of an econometric model of metropolitan growth" *Journal of Urban Economics* **1** 250-267

Engle R F, Fisher F M, Harris J R, Rothenberg J G, 1972 "An economic simulation model of intra-metropolitan housing location: housing, business, transportation and local government" *American Economic Review* Papers **62**, 87-97

Evans A W, 1972 "The pure theory of city size in an industrial economy" *Urban Studies* **9** 49-77

Evans A W, 1974 *The Economics of Residential Location* (Macmillan, London)

Evans A W, 1975 "Rents and housing in the theory of urban growth" *Journal of Regional Science* **15** 113-125

Fales R L, Moses L N, 1972 "Land use theory and the spatial structure of the nineteenth-century city" *Papers and Proceedings of the Regional Science Association* **28** 49-80

Farhi A, 1973 "Urban economic growth and conflicts: a theoretical approach" *Papers of the Regional Science Association* **31** 95-124

Firey W, 1947 *Land Use in Central Boston* (Harvard University Press, Cambridge, Mass.)

Fisch O, 1974a "Optimal land allocation and city size" Department of City and Regional Planning, Ohio State University, Columbus (mimeo)

Fisch O, 1974b "Externalities, urban rent and population density functions: the case of air pollution" Department of City and Regional Planning, Ohio State University, Columbus (mimeo)

Fisch O, 1974c "Impact analysis on optimal densities and optimal city size" *Journal of Regional Science* **14** 233-246

- Fisch O, 1975 "Variable residential density: optimum traffic density and city size" *Geographical Analysis* **7** 107-119
  Fisch O, 1976a "Spatial equilibrium with local public goods: urban land rent, optimal city size and the Tiebout hypothesis" in *Mathematical Land Use Theory* Ed. G J Papageorgiou (Lexington Books, Lexington, Mass.) pp 177-193
  Fisch O, 1976b "Optimal city size, land tenure and the economic theory of clubs" *Regional Science and Urban Economics* **6** 33-44
  Forrester J W, 1969 *Urban Dynamics* (MIT Press, Cambridge, Mass.)
- Forsund F R, 1972 "Allocation in space and environmental pollution" *Swedish Journal of Economics* **74** 19-35
- Freeman III M A, 1971 "Air pollution and property values: a methodological comment" *Review of Economics and Statistics* **53** 415-416
  Freeman III M A, 1974 "On the estimation of air pollution control benefits from land values studies" *Journal of Environmental Economics and Management* **1** 74-83
- Gannon C A, 1973 "Intraurban location and interestablishment linkages" *Geographical Analysis* **5** 214-244
  Garin R A, 1966 "A matrix formulation of the Lowry model for intrametropolitan activity allocation" *Journal of the American Institute of Planners* **32** 361-364
- Getz M, 1975 "A model of the impact of transportation investments on land rents" *Journal of Public Economics* **4** 57-74
- Glickman N J, 1974 "Son of 'Specification of regional econometric models'" *Papers, Regional Science Association* **32** 155-177
  Gohman V M, Karpov L N, 1972 "Growth poles and growth centres" in *Growth Poles and Growth Centres in Regional Planning* Ed. A Kuklinski (Mouton, The Hague)
  Goldberg M A, 1970 "An economic model of intrametropolitan industrial location" *Journal of Regional Science* **10** 75-79
- Goldberg V P, Kriesel K M, 1971 "Urban form and the allocation of land to streets" Research Report number 22, Institute of Government Affairs, University of California, Davis
  Goldner W, 1971 "The Lowry model heritage" *Journal of the American Institute of Planners* **37** 100-110
- Goldstein G S, Moses L N, 1973 "A survey of urban economics" *Journal of Economic Literature* **11** 471-495
- Goldstein G S, Moses L N, 1975 "Interdependence and the location of economic activities" *Journal of Urban Economics* **2** 63-84
  Gordon G, 1971 *Status Areas in Edinburgh* PhD thesis, University of Edinburgh, Edinburgh, Scotland
  Guigou J L, 1972 *Théorie Économique et Transformation de l'Espace Agricole* (Gauthier-Villars, Paris)
  Gutnov A, and others, 1970 *The Ideal Communist City* (Braziller, New York)
  Haig R M, 1926 "Towards an understanding of the metropolis" *Quarterly Journal of Economics* **40** 197-208
  Hall P G, 1966 *von Thünen's Isolated State* (Pergamon Press, Oxford)
- Hanel P, 1973 "Does circular causation lead to increased growth rate?" (mimeo)
- Haring J E, Slobko T, Chapman J, 1976 "The impact of alternative transportation systems on urban structure" *Journal of Urban Economics* **3** 14-30
- Harris B, and others, 1966 *Research on an Equilibrium Model of Metropolitan Housing and Locational Choice* Interim Report, Department of City and Regional Planning, University of Pennsylvania, Philadelphia
  Harris C D, Ullman E L, 1945 "The nature of cities" *Annals of the American Academy of Political and Social Science* **242** 7-17

Harris J R, Wheeler D, 1972 "Agglomeration economies: theory and measurement" CP 6, in *Papers from the Urban Economics Conference, Keele 1971* (Centre for Environmental Studies, London)
• Harrison D Jr, Kain J F, 1974 "Cumulative urban growth and urban density functions" *Journal of Urban Economics* 1 61-98
Hartwick J M, 1974 "Price sustainability of location assignments" *Journal of Urban Economics* 1 147-161
• Hartwick J M, 1976 "Intermediate goods and the spatial integration of land uses" *Regional Science and Urban Economics* 6 127-145
Hartwick J M, Schweitzer U, Varaiya P, 1976 "Comparative statics of a residential economy with several classes" in *Essays in Mathematical Land Use Theory* Ed. G J Papageorgiou (Lexington Books, D C Heath, Lexington, Mass.) pp 55-78
Hartwick P G, Hartwick J M, 1972 "An analysis of an urban thoroughfare" *Environment and Planning* 4 193-204
Hartwick P G, Hartwick J M, 1974 "Efficient resource allocation in a multinucleated city with intermediate goods" *Quarterly Journal of Economics* 88 340-352
Harvey D, 1973 *Social Justice and the City* (Edward Arnold, London)
Harvey D, 1974 "Class-monopoly rent, finance capital and the urban revolution" *Regional Studies* 8 239-255
Harvey D, 1975 *The Political Economy of Urbanization in Advanced Capitalist Societies—the Case of the United States* (Johns Hopkins Centre for Metropolitan Planning and Research, Baltimore)
• Haugen R A, Heins A J, 1969 "A market separation theory of rent differentials in metropolitan areas" *Quarterly Journal of Economics* 83 660-673
Henderson J V, 1974a "Road congestion: a reconsideration of pricing theory" *Journal of Urban Economics* 1 346-365
Henderson J V, 1974b "Optimum city size: the external diseconomy question" *Journal of Political Economy* 82 373-388
• Henderson J V, 1975 "Congestion and the optimum city size" *Journal of Urban Economics* 2 48-62
Herbert J D, Stevens B H, 1960 "A model of the distribution of residential activity in urban areas" *Journal of Regional Science* 2 21-36
• Hilton G, 1976 "What did we give up with the big red cars?" in *Transportation Alternatives for Southern California* Eds P Gordon, R D Eckert (Institute for Public Policy Research, Center for Public Affairs, University of Southern California, Los Angeles) pp 86-97
Hoch I, 1969 "The three-dimensional city: contained urban space" in *The Quality of the Urban Environment* Ed. H E Perloff (Resources for the Future, Washington, DC) pp 75-138
Hoch I, 1972 "Income and city size" *Urban Studies* 9 299-328
• Hoch I, 1974 "Interurban differences in the quality of life" in *Transport and the Urban Environment* Eds J G Rothenberg, I G Heggie (Macmillan, London) pp 54-90
Hochman O, Pines D, 1971 "Competitive equilibrium of transportation and housing in the residential ring of an urban area" *Environment and Planning* 3 51-62
• Hochman O, Pines D, 1972 "Note on land use in a long narrow city" *Journal of Economic Theory* 5 540-541
Hochman O, Pines D, 1973 "Dynamic aspects of land use patterns in a growing city", WP 17, Centre for Urban and Regional Studies, Tel Aviv University, Tel Aviv
Hoselitz B F, 1954-1955 "Generative and parasitic cities" *Economic Development and Cultural Change* 3 278-294
Hoyt H, 1939 *Structure and Growth of Residential Neighborhoods in American Cities* (Federal Housing Administration, Washington, DC)

Hoyt H, 1951 "Is city growth controlled by mathematics or physical laws?" *Land Economics* **27** 259-262

Hurd R M, 1903 *Principles of City Land Values* (The Record and Guide, New York)

Ingram G K, Kain J F, Ginn J R, 1972 *The Detroit Prototype of the NBER Urban Simulation Model* (Columbia University Press, New York, for the National Bureau of Economic Research)

Isard W, 1956 *Location and Space-Economy* (MIT Press, Cambridge, Mass.)

• Ives R, Lloyd G, Sawers L, 1972 "Mass transit and the power elite" *Review of Radical Political Economics* **4** 68-77

Kadanoff L P, 1971 "From simulation model to public policy: an examination of Forrester's *Urban Dynamics*" *Simulation* **16** 261-268

Kadanoff L P, 1972 "From simulation model to public policy" *American Scientist* **60** 74-79

• Kanemoto Y, 1974 "A note on a concealed nonconvexity in urban residential models" (mimeo)

Kanemoto Y, 1975 "Congestion and cost-benefit analysis in cities" *Journal of Urban Economics* **2** 246-264

• Kanemoto Y, 1976 "Optimum, market and second best land use patterns in a von Thünen city with congestion" *Regional Science and Urban Economics* **6** 23-32

Kemper P, Schmenner R, 1974 "The density gradient for manufacturing" *Journal of Urban Economics* **1** 410-427

Kendrick D, 1972 "Numerical models for urban planning" *Swedish Journal of Economics* **74** 45-67

• King A T, Mieszkowski P, 1973 "Racial discrimination, segregation and the price of housing" *Journal of Political Economy* **81** 590-606

Kirwan R S, Ball M J, 1974 "The microeconomic analysis of a local housing market" CP9, in *Papers from the Urban Economics Conference, 1973, Volume 1* (Centre for Environmental Studies, London) pp 115-199

Koopmans T C, 1957 *Three Essays on the State of Economic Science* (McGraw-Hill, New York)

Koopmans T C, Beckmann M J, 1957 "Assignment problems in the location of economic activities" *Econometrica* **25** 53-76

Kraus M, 1974 "Land use in a circular city" *Journal of Economic Theory* **8** 440-457

Kumar-Misir L M, 1974 *Regional Economic Growth in Canada: An Urban-Rural Functional Area Analysis* MA thesis, University of Ottawa, Ottawa, Canada

• Lapham V, 1971 "Do blacks pay more for housing?" *Journal of Political Economy* **79** 1244-1257

• Latham R F, Yeates M H, 1970 "Population density growth in metropolitan Toronto" *Geographical Analysis* **2** 177-185

Lave L B, 1970 "Congestion and urban location" *Papers and Proceedings of Regional Science Association* **25** 133-149

Lave L B, 1974 "Urban externalities" CP9, in *Papers from the Urban Economics Conference 1973, Volume 1* (Centre for Environmental Studies, London) pp 37-95

• Lee D B, 1974 "Requiem for large-scale models" *Journal of the American Institute of Planners* **40** 163-178

• Lee D B, Averaus C P, 1973 "Land use and transportation in basic theory" *Environment and Planning* **5** 491-502

Lefebvre H, 1970 *La Révolution Urbaine* (Gallimard, Paris)

Lefebvre H, 1972 *La Pensée Marxiste et la Ville* (Casterman, Paris)

Legey L, Ripper M, Varaiya P, 1973 "Effects of congestion on the shape of a city" *Journal of Economic Theory* **6** 162-179

Levhari D, Oron Y, Pines D, 1972 "In favour of lottery in cases of non-convexities" WP 10, Centre for Urban Studies, Tel Aviv University, Tel Aviv

- Lind R C, 1973 "Spatial equilibrium, the theory of rents and public program benefits" *Quarterly Journal of Economics* **87** 188-207
  Livesey D A, 1973 "Optimum city size: a minimum congestion cost approach" *Journal of Economic Theory* **6** 144-161
  Livesey D A, 1976 "Optimum and market land rents in a CBD city" in *Essays in Mathematical Land Use Theory* Ed. G J Papageorgiou (Lexington Books, D C Heath, Lexington, Mass.) pp 215-227
  Long W H, 1971 "Demand in space: some neglected aspects" *Papers, Regional Science Association* **27** 54-60
  Lösch A, 1954 *The Economics of Location* (Yale University Press, New Haven, Conn.)
  Lowry I S, 1964 *A Model of Metropolis* RM-4035-RC, Rand Corporation, Santa Monica, California
  MacKinnon J, 1974 "Urban general equilibrium models and simplicial search algorithms" *Journal of Urban Economics* **1** 161-183
- Martin R C, 1973 "Spatial distribution of population: cities and suburbs" *Journal of Regional Science* **13** 269-278
  Marx K, 1967 edition *Capital* 3 volumes (International Publishers, New York)
- McDougall G, 1975 "Individual optimization in an urban area: a formalization of the Tiebout hypothesis" *American Economist* **18** 91-95
- McKean R N, 1973 "An outsider looks at urban economics" *Urban Studies* **10** 19-37
  Meadows D H, Meadows D L, Randers J, Behrens W W III, 1972 *The Limits to Growth* (New American Library, New York)
  Mills E S, 1967 "An aggregative model of resource allocation in a metropolitan area" *American Economic Review* Papers **57**, 197-210
  Mills E S, 1970 "The efficiency of spatial competition" *Papers and Proceedings of the Regional Science Association* **25** 71-82
  Mills E S, 1972a *Studies in the Structure of the Urban Economy* (Johns Hopkins University Press, Baltimore)
  Mills E S, 1972b "Markets and efficient resource allocation in urban areas" *Swedish Journal of Economics* **74** 100-113
- Mills E S, 1974 "Mathematical models for urban planning" Department of Economics, Princeton University, Princeton
  Mills E S, Ferranti D M de, 1971 "Market choices and optimum city size" *American Economic Review* **61** 360-365
  Mills E S, MacKinnon J, 1973 "Notes on the new urban economics" *Bell Journal of Economics and Management Science* **4** 593-601
  Mirrlees J A, 1972 "The optimum town" *Swedish Journal of Economics* **74** 114-135
- Miyao T, 1975 "Dynamics and comparative statics in the theory of residential location" *Journal of Economic Theory* **11** 133-146
- Mogridge M J H, 1969 "Some factors influencing the income distribution of households within a city region" in *London Papers in Regional Science 1. Studies in Regional Science* Ed. A J Scott (Pion, London) pp 117-141
  Mogridge M J H, 1974 "Some thoughts on the economics of intra-urban spatial location of homes, worker-residences and workplaces" CP9, in *Papers from the Urban Economics Conference, 1973, Volume 1* (Centre for Environmental Studies, London) pp 259-301
  Mohring H, Harwitz M, 1962 *Highway Benefits: An analytical Framework* (Northwestern University Press, Evanston, Ill.)
  Montesano A, 1972 "A restatement of Beckmann's model on the distribution of urban rent and residential density" *Journal of Economic Theory* **4** 329-354
  Morrill R L, 1968 "Waves of spatial diffusion" *Journal of Regional Science* **8** 1-18
  Morrill R L, 1970 "The shape of diffusion in space and time" *Economic Geography* **46** 259-268

Moses L N, 1962 "Towards a theory of intra-urban wage differentials and their influence on travel patterns" *Papers and Proceedings of the Regional Science Association* 9 53-63

Moses L N, Williamson H F Jr, 1967 "The location of economic activity in cities" *American Economic Review* Papers 57, 211-238

• Muth R F, 1960 "The demand for nonfarm housing in *The Demand for Durable Goods* Ed. A C Harberger (Chicago University Press, Chicago)

Muth R F, 1961a "The spatial structure of the housing market" *Papers and Proceedings of the Regional Science Association* 7 207-220

Muth R F, 1961b "Economic change and urban-rural land conversions" *Econometrica* 29 1-23

• Muth R F, 1967 "Comment on Moses and Williamson" *American Economic Review* Papers 57, 239-241

• Muth R F, 1968 "Urban residential land and housing markets" in *Issues in Urban Economics* Eds H S Perloff, L Wingo (Johns Hopkins University Press, Baltimore)

Muth R F, 1969 *Cities and Housing* (Chicago University Press, Chicago)

• Muth R F, 1971 "The derived demand for urban residential land" *Urban Studies* 8 243-254

Muth R F, 1972a "A vintage model of the housing stock" *Papers, Regional Science Association* 30 141-156

Muth R F, 1972b "Comment on Engle et al." *American Economic Review* Papers 62, 98-99

• Muth R F, 1974 "Moving costs and housing expenditures" *Journal of Urban Economics* 1 108-125

• Muth R F, 1975 "Numerical solution of urban residential land-use models" *Journal of Urban Economics* 2 307-332

Muth R F, 1976 "A vintage model with housing production" in *Essays in Mathematical Land Use Theory* Ed. G J Papageorgiou (Lexington Books, D C Heath, Lexington, Mass.) pp 245-259

• Nelson R H, 1971 *The Theory of Residential Location* PhD thesis, Princeton University, Princeton, NJ

• Nelson R H, 1973 "Accessibility and rent: applying Becker's 'time price' concept to the theory of residential location" *Urban Studies* 10 83-86

Niedercorn J H, 1971 "A negative exponential model of urban land use densities and its implications for metropolitan development" *Journal of Regional Science* 11 317-326

Niedercorn J H, 1974 "A model for estimating trade-offs between agricultural and urban land use" RP 12, Department of Economics, University of Southern California, Los Angeles

Nourse H G, 1967 "The effect of air pollution on property values" *Land Economics* 43 181-189

Oates W E, 1969 "The effects of property taxes and local public spending on property values: an empirical study of the tax capitalization and the Tiebout hypothesis" *Journal of Political Economy* 77 957-971

Oates W E, Howrey E P, Baumol W J, 1971 "The analysis of public policy in dynamic urban models" *Journal of Political Economy* 79 142-153

• O'Connor J, 1972 "The fiscal crisis of the state" in *Readings in Urban Economics* Eds M Edel, J G Rothenberg (Macmillan, New York) pp 590-602

• Ohls J C, 1975 "Public policy toward low income housing and filtering in housing markets" *Journal of Urban Economics* 2 144-171

• Ohls J C, Pines D, 1975 "Discontinuous urban development and economic efficiency" *Land Economics* 51 52-68

Ohls J C, Weisberg R C, White M J, 1974 "The effects of zoning and land value" *Journal of Urban Economics* **1** 428-444
• Ohls J C, Weisberg R C, White M J, 1976 "Welfare effects in alternative models of zoning" *Journal of Urban Economics* **3** 95-96
• Olsen E O, 1969 "A competitive theory of the housing market" *American Economic Review* **59** 612-621
• Oron Y, 1972 *Optimal Resource Allocation vs. Competitive Resource Allocation in a Mono-Center Urban Form* PhD thesis, Tel Aviv University, Tel Aviv
Oron Y, Pines D, 1975 "The effect of efficient pricing of air pollution on intraurban land-use patterns" *Environment and Planning A* **7** 293-299
Oron Y, Pines D, Sheshinski E, 1973 "Optimum vs. equilibrium land use patterns and congestion toll" *Bell Journal of Economics and Management Science* **4** 619-636
Oron Y, Pines D, Sheshinski E, 1974 "The effect of nuisances associated with urban traffic on suburbanization and land values" *Journal of Urban Economics* **1** 382-394
• Papageorgiou G J, 1971a "A generalization of the density gradient concept" *Geographical Analysis* **3** 121-127
• Papageorgiou G J, 1971b "A theoretical evaluation of the existing population density gradient functions" *Economic Geography* **47** 21-26
• Papageorgiou G J, 1971c "The population density and rent distribution models within a multicentre framework" *Environment and Planning* **3** 267-282
• Papageorgiou G J, 1973 "The impact of the environment upon the spatial distribution of population and land values" *Economic Geography* **49** 251-256
Papageorgiou G J, 1974 "Spatial equilibrium within a hierarchy of centres with distributed incomes" Department of Geography, McMaster University, Hamilton, Ontario
• Papageorgiou G J, 1975 "Urban residential analysis" Department of Geography, McMaster University, Hamilton, Ontario
Papageorgiou G J (Ed.), 1976a *Essays in Mathematical Land Use Theory* (Lexington Books, D C Heath, Lexington, Mass.)
Papageorgiou G J, 1976b "On spatial consumer equilibrium" in *Essays in Mathematical Land Use Theory* Ed. G J Papageorgiou (Lexington Books, D C Heath, Lexington, Mass.) pp 145-176
Papageorgiou G J, 1976c "Urban residential analysis: 1. Spatial consumer behaviour" *Environment and Planning A* **8** 423-442
Papageorgiou G J, 1976d "Urban residential analysis: 2. Spatial consumer equilibrium" *Environment and Planning A* **8** 489-506
Papageorgiou G J, 1976e "On issues of urban policy" paper presented at the North American Conference of the Regional Science Association, Toronto, Canada
Papageorgiou G J, Casetti E, 1971 "Spatial equilibrium residential land values in a multicentric setting" *Journal of Regional Science* **3**(3) 385-389
• Papageorgiou G J, Mullaly H, 1974 "Urban residential structure" Department of Geography, McMaster University, Hamilton, Ontario (mimeo)
Parr J B, 1970 "Models of city size in an urban system" *Papers and Proceedings of the Regional Science Association* **25** 221-253
• Pines D, 1970 "The exponential density function: a comment" *Journal of Regional Science* **10** 107-110
Pines D, 1972 "The equilibrium utility level and city size: a comment" *Economic Geography* **48** 439-443
• Pines D, 1974 "A note on the relationship between public programs benefits and its effect on equilibrium rent" Department of Economics, Princeton University, Princeton, NJ (mimeo)
Pines D, 1975 "On the spatial distribution of households according to income" *Economic Geography* **51** 142-149

Pines D, 1976 "Dynamic aspects of land use pattern in a growing city" in *Essays in Mathematical Land Use Theory* Ed. G J Papageorgiou (Lexington Books, D C Heath, Lexington, Mass.) pp 229-242
• Pines D, Weiss Y, 1976 "Land improvement projects and land values" *Journal of Urban Economics* 3 1-13
Pióro Z, 1972 "Growth poles and growth centres theory as applied to settlement development in Tanzania" in *Growth Poles and Growth Centres in Regional Planning* Ed. A Kuklinski (Mouton, The Hague) pp 169-194
Polinsky A M, Rubinfeld D L, 1974 "The long-run incidence of a residential property tax and local public services " DP 354, Harvard Institute of Economic Research, Harvard University, Cambridge, Mass.
Polinsky A M, Shavell S, 1973 "Amenities and property values in a general equilibrium model of an urban area" WP 1207-5, The Urban Institute, Washington, DC
• Polinsky A M, Shavell S, 1975 "The air pollution and property value debate" *Review of Economics and Statistics* 57 100-104
• Pollakowski H O, 1973 "The effects of property taxes and local public spending on property values: a comment and further results" *Journal of Political Economy* 81 994-1003
• Rabenau B von, 1976 "Optimal growth of a factory town" *Journal of Urban Economics* 3 97-112
Rashevsky N, 1951 *Mathematical Biology of Social Behavior* (Chicago University Press, Chicago)
• Rashevsky N, 1967 "Physics, biology and sociology: II. Suggestions for a synthesis" *Bulletin of Mathematical Biophysics* 29 643-648
• Rashevsky N, 1970 "A remark on the course of development of organismic sets" *Bulletin of Mathematical Biophysics* 32 79-81
Ratcliff R U, 1949 *Urban Land Economics* (McGraw-Hill, New York)
Rawls J, 1971 *A Theory of Justice* (Harvard University Press, Cambridge, Mass.)
Richardson H W, 1972 *Input-Output and Regional Economics* (Weidenfeld and Nicolson, London)
Richardson H W, 1973a "A comment on some uses of mathematical models in urban economics" *Urban Studies* 10 259-270
Richardson H W, 1973b *The Economics of Urban Size* (Saxon House, Farnborough, Hants)
Richardson H W, 1973c *Regional Growth Theory* (Macmillan, London)
Richardson H W, 1974 "Agglomeration potential: a generalization of the income potential concept" *Journal of Regional Science* 14 325-336
• Richardson H W, 1975a "Discontinuous densities, urban spatial structure and growth: a new approach" *Land Economics* 51 305-315
• Richardson H W, 1976 "Relevance of mathematical land use theory to applications" in *Essays in Mathematical Land Use Theory* Ed. G J Papageorgiou (Lexington Books, D C Heath, Lexington, Mass.) pp 9-22
Richardson H W, 1977 "On the possibility of positive rent gradients" *Journal of Urban Economics* 4 (forthcoming)
Richardson H W, Vipond M J, Furbey R A, 1974a "The determinants of urban house prices" *Urban Studies* 11 190-199
Richardson H W, Vipond M J, Furbey R A, 1974b "Dynamic tests of Hoyt's spatial model" *Town Planning Review* 45 401-414
• Richardson H W, Vipond M J, Furbey R A, 1975 *Housing and Urban Spatial Structure: A Case Study* (Saxon House, Farnborough, Hants)
Ridker R G, Henning J A, 1967 "The determinants of residential property values with special reference to air pollution" *Review of Economics and Statistics* 49 246-257

- Riley J G, 1972 *Optimal Towns* PhD thesis, Massachusetts Institute of Technology, Cambridge, Mass.)
  Riley J G, 1973 "Gammaville: an optimum town" *Journal of Economic Theory* **6** 471-482
  Riley J G, 1974 "Optimal residential density and road transportation" *Journal of Urban Economics* **1** 230-250
  Ripper M, Varaiya P, 1974 "An optimizing model of urban development" *Environment and Planning A* **6** 149-168
- Robson A J, 1976 "Cost-benefit analysis and the use of urban land for transportation" *Journal of Urban Economics* **3** 180-191
  Rodwin L, 1961 *Housing and Economic Progress: A Study of the Housing Experiences of Boston's Middle Income Families* (Harvard University Press and the Technology Press, Cambridge, Mass.)
- Rogers A, 1966 "A note on the Garin-Lowry model" *Journal of the American Institute of Planners* **32** 364-366
- Rose-Ackerman S, 1974 "On the use of urban economics in cost-benefit analysis" (mimeo)
  Rose-Ackerman S, 1975 "Racism and urban structure" *Journal of Urban Economics* **2** 85-103
- Rothenberg J G, 1973 "Discussion on 'congestion and optimum city size'" *American Economic Review* Papers 63, 67-70
- Rothenberg J G, 1974 "Problems in the modelling of urban development: a review article on *Urban Dynamics* by Jay W. Forrester" *Journal of Urban Economics* **1** 1-20
  Sawers L, 1975 "Urban form and the mode of production" *Review of Radical Political Economics* **7** 52-68
- Schelling T C, 1969 *Models of Segregation* RM-6014-RC, Rand Corporation, Santa Monica, California
  Schelling T C, 1971 "On the ecology of micromotives" *The Public Interest* **25** 59-98
- Schelling T C, 1972 "Neighborhood tipping" in *Racial Discrimination in Economic Life* Ed. A Pascal (Lexington Books, D C Heath, Lexington, Mass.)
  Schultz T W, 1953 *The Economic Organization of Agriculture* (McGraw-Hill, New York)
  Scott A J, 1975a "Land rent, land use and transport: a study in the geographical foundations of political economy" Research Report 27, University of Toronto-York University Joint Program in Transportation,
  Scott A J, 1975b "Transport, residential space, and the urbanization process" Research Report 28, University of Toronto-York University Joint Program in Transportation,
  Scott A J, 1976a "Land use and commodity production: an elementary synthesis of the von Thünen and Sraffa models" *Regional Science and Urban Economics* **6** 147-160
  Scott A J, 1976b "Land and land rent: an interpretative review of the French literature" *Progress in Geography* (forthcoming)
  Segal D, 1974 "Are there returns to scale in city size?" Resources for the Future, Washington, DC, mimeo
- Senior M L, 1974 "Approaches to residential location modelling 2: urban economic models and some recent developments (a review)" *Environment and Planning A* **6** 369-409
  Senior M L, Wilson A G, 1974 "Disaggregated residential location models: some tests and further theoretical developments" in *London Papers in Regional Science 4. Space-Time Concepts in Urban and Regional Models* Ed. E L Cripps (Pion, London) pp 141-172

References

Shefer D, 1973 "Localization economics in SMSAs: a production function analysis" *Journal of Regional Science* **13** 55-64
• Sherratt G G, 1960 "A model for general urban growth" *Management Sciences, Models and Techniques* **2** 147-159
Sheshinski E, 1973 "Congestion and the optimum city size" *American Economic Review* **63** 61-66
• Smolensky E, Becker S, Molotch H, 1968 "The prisoner's dilemma and ghetto expansion" *Land Economics* **44** 420
Solow R M, 1956 "A contribution to the theory of economic growth" *Quarterly Journal of Economics* **52** 65-94
Solow R M, 1972 "Congestion, density and the use of land in transportation" *Swedish Journal of Economics* **74** 161-173
Solow R M, 1973a "Congestion costs and the use of land for streets" *Bell Journal of Economics and Management Science* **4** 602-618
Solow R M, 1973b "Rejoinder to Richardson—I" *Urban Studies* **10** 267
Solow R M, 1973c "On equilibrium models of urban location" in *Essays in Modern Economics* Ed. J M Parkin (Longmans, London) pp 2-16
Solow R M, Vickrey W S, 1971 "Land use in a long narrow city" *Journal of Economic Theory* **3** 430-447
Starrett D A, 1972 "On the optimal degree of increasing returns" DP 230, Harvard University of Economic Research, Harvard University, Cambridge, Mass.
• Starrett D A, 1974 "Principles of optimal location in a large homogeneous area" *Journal of Economic Theory* **9** 418-448
• Stern N H, 1972 "The optimal size of market areas" *Journal of Economic Theory* **4** 154-173
Stern N H, 1973 "Homogeneous utility functions and equality in 'the optimum town'" *Swedish Journal of Economics* **75** 204-207
Stewart C T Jr, 1958 "The size and spacing of cities" *Geographical Review* **48** 222-245
Strotz R H, 1965 "Urban transportation parables" in *The Public Economy of Urban Communities* Ed. J Margolis (Johns Hopkins University Press, Baltimore) pp 127-169
• Strotz R H, 1968 "The use of land rent changes to measure the welfare benefits of land improvement" in *The New Economics of Regulated Industries: Rate Making in a Dynamic Economy* Ed. J E Haring (Occidental College, Los Angeles Economic Research Center, Los Angeles) pp 174-186
Strotz R H, Wright C, 1975 "Spatial adaptation to urban air pollution" *Journal of Urban Economics* **2** 212-222
• Stucker J P, 1975 "Transport improvements, commuting costs and residential location" *Journal of Urban Economics* **2** 123-143
• Stull W J, 1973 "A note on residential bid price curves" *Journal of Regional Science* **13** 107-113
• Stull W J, 1974 "Land use and zoning in an urban economy" *American Economic Review* **44** 337-347
• Sveikauskas L, 1975 "The productivity of cities" *Quarterly Journal of Economics* **89** 393-413
Swan T W, 1956 "Economic growth and capital accumulation" *Economic Record* **32** 334-361
Swanson J A, Smith K R, Williamson J G, 1974 "The size distribution of cities and optimal city size" *Journal of Urban Economics* **1** 395-409
• Sweeney J L, 1974 "A commodity hierarchy model of the rental housing market" *Journal of Urban Economics* **1** 288-324
• Taylor P J, 1971 "Distance transformation and distance decay functions" *Geographical Analysis* **3** 221-238

Thünen J H von, 1966 *Der Isolierte Staat in Beziehung auf Nationalökonomie und Landwirtschaft* (Gustav Fischer, Stuttgart) reprint of 1826
Tiebout C M, 1956 "A pure theory of local expenditures" *Journal of Political Economy* 64 416-424
Tinbergen J, 1968 "The hierarchy model of the size distribution of centres" *Papers and Proceedings of the Regional Science Association* 20 65-68
Tisdell C, 1974 "The theory of optimal city-size: elementary speculations about analysis and policy" OP 8, Department of Economics, University of Newcastle, New South Wales, Australia
Tolley G S, 1974 "The welfare economics of city bigness" *Journal of Urban Economics* 1 324-346
• Ullman E L, 1968 "The nature of cities reconsidered" *Papers and Proceedings of the Regional Science Association* 9 7-23
• Varaiya P, Artle R, 1972 "Locational implications of transaction costs" *Swedish Journal of Economics* 74 174-183
Vernon R, 1960 *Metropolis 1985* (Harvard University Press, Cambridge, Mass.)
Vickrey W S, 1965 "Pricing as a tool in co-ordination of local transportation" in *Transportation Economics* Ed. J R Meyer (National Bureau for Economic Research, New York)
Vickrey W S, 1969 "Congestion theory and transport investment" *American Economic Review* 59 251-260
Wabe J S, 1971 "A study of house prices as a means of establishing the value of journey time, the rate of time preference and the valuation of some aspects of environment in the London metropolitan region" *Applied Economics* 3 247-256
• Webber M J, 1973 "Equilibrium of location in an isolated state" *Environment and Planning* 5 751-759
• Webber M J, 1976 "The meaning of entropy maximizing models" in *Essays in Mathematical Land Use Theory* Ed. G J Papageorgiou (Lexington Books, D C Heath, Lexington, Mass.) pp 277-292
Weiss H K, 1961 "The distribution of urban population and an application to a servicing problem" *Operations Research* 9 860-874
• Wheaton W C, 1972 *Income and Urban Location* PhD thesis, University of Pennsylvania, Philadelphia
Wheaton W C, 1974a "Linear programming and locational equilibrium: the Herbert-Stevens model revisited" *Journal of Urban Economics* 1 278-287
Wheaton W C, 1974b "A comparative static analysis of urban spatial structure" *Journal of Economic Theory* 9 223-237
• Wheaton W C, 1976 "On the optimal distribution of income among cities" *Journal of Urban Economics* 3 31-44
• White M J, 1975 "The effect of zoning on the size of metropolitan areas" *Journal of Urban Economics* 2 279-290
Wieand K F Jr, 1973 "Air pollution and property values: a study of the St. Louis area" *Journal of Regional Science* 13 91-95
• Wieand K F Jr, Muth R F, 1972 "A note on the variation of land values with distance from the CBD in St. Louis" *Journal of Regional Science* 12 469-473
• Wilkins C A, 1968 "Some points in the methodology of urban population distributions" *Operations Research* 16 1-9
Wilkinson R K, 1972 "The determinants of relative house prices" CP6, in *Papers from the Urban Economics Conference, Keele, 1971* (Centre for Environmental Studies, London)
• Wilson A G, 1967 "A statistical theory of spatial trip distribution models" *Transportation Research* 1 253-269
Wilson A G, 1970 *Entropy in Urban and Regional Modelling* (Pion, London)

References

• Wilson A G, 1971 "A family of spatial interaction models, and associated developments" *Environment and Planning* **3** 1-32

Wilson A G, 1974 *Urban and Regional Models in Geography and Planning* (John Wiley, Chichester)

Wingo L, 1961a *Transportation and Urban Land* (Resources for the Future, Washington, DC)

Wingo L, 1961b "An economic model of the utilization of land for residential purposes" *Papers and Proceedings of the Regional Science Association* **7** 191-205

• Wright C, 1971 "Residential location in a three-dimensional city" *Journal of Political Economy* **79** 1378-1387

Yamada H, 1972 "On the theory of residential location: accessibility, space, leisure and environmental quality" *Papers of the Regional Science Association* **29** 125-135

• Yellin J, 1974 "Urban population distribution, family income and social prejudice" *Journal of Urban Economics* **1** 21-47

Zipf G K, 1949 *Human Behavior and the Principle of Least Effort* (Addison-Wesley, Cambridge, Mass.)

Name index

Alao N 33, 69, 71, 72, 84, 89, 238
Alonso W 1, 3, 6, 14, 16, 18, 25, 105, 111, 168, 213, 219
Amson J C 41, 113, 159
Anas A 1, 56, 124, 125, 126, 165, 242
Anderson M 225
Anderson R 60
Angel S 86, 116, 237
Arrow K J 40, 74
Artle R 6, 40, 75, 77, 238, 239

Bailey M J 22
Ball M J 120
Banfield E C 230, 231, 232, 233
Barr J L 25, 134
Bateman W 232
Batty M 181, 183
Baumol W J 178, 181, 209, 240, 241
Becker G S 106
Beckmann M J 2, 4, 5, 6, 18, 31, 40, 66, 67, 78, 80, 111, 149, 160, 238
Ben-Shahar H 124
Blumenfeld H 212
Borukhov E 25, 35
Böventer E G von 141, 147
Bradford D F 180, 198, 241
Burgess E W 2, 10, 222
Bussière R 130, 243

Capozza D R 36, 58, 59, 60, 80, 81, 136, 237
Casetti E 32, 52, 53, 89, 95, 97, 111, 238
Christaller W 93, 94, 110, 146
Clark C 2, 21, 158
Cordey-Hayes M 210
Crocker T 60
Curry L 146

Davies G W 36, 58, 59
Davis E 144
Debreu G 40
Delson J K 4
Dendrinos P 1
Deutsch K W 92
Devletoglou N E 30, 122
Dhrymes P J 74
Dixit A 33, 50, 51, 54, 55, 58, 68, 69, 70, 71, 72, 89, 90, 93, 238

Edel M 25, 72, 134, 144, 217, 223
Ellickson B 142
Engels F 214, 222, 224, 225, 226, 227, 228, 229
Engle R F 194
Evans A W 106, 127, 128, 141, 243

Fales R L 32, 71, 80, 81, 83, 237, 242
Farhi A 19, 168, 218
Ferranti D M de 33, 55, 58, 69, 70, 125
Firey W 14
Fisch O 37, 55, 62, 63
Forrester J W 203, 204, 205, 208, 210, 211, 241
Frank A G 227
Freeman A M III 60, 61

Garin R A 181
Gohman V M 229
Goldberg M A 83
Goldner W 181, 240
Gordon G 13
Guigou J L 6
Gutnov A 214, 229

Haig R M 2, 23
Hall P G 2, 6
Harris J R 72
Hartwick J M 1, 40, 41, 58, 95, 103, 149, 174, 238
Hartwick P G 1, 41, 58, 95, 149, 174, 238
Harvey D 168, 214, 215, 216, 217, 220, 221, 222
Harwitz M 2
Henderson J V 57
Henning J A 60, 61, 239
Herbert J D 37, 168, 172, 173
Hilton G 222
Hobson J A 227
Hoch I 56, 64, 140
Hochman H M 232
Hochman O 50, 55, 125, 238
Hoselitz B F 227
Hoyt H 2, 6, 11, 12, 13, 14, 119
Hurd R M 2, 6, 10
Hyman G M 86, 116, 237

Ingram G K 37, 185, 187
Isard W 6, 92

Kadanoff L P 211
Kain J F 185
Kanemoto Y 58
Karpov L N 229
Kelejian H 180, 241
Kemper P 83, 237
Kendrick D 181
Kirwan R S 120
Koopmans T C 40, 68
Kraus M 36, 56
Kumar-Misir L M 181

Lave L B 32, 58, 70, 90, 98, 99, 238
Lefebvre H 215, 230
Legey L 56, 58
Lenin V 227
Levhari D 54
Livesey D A 32, 33, 54, 55, 58, 69, 70, 71, 84, 89, 90, 238
Long W H 110
Lösch A 93, 94, 110, 145, 146, 149
Lowry I S 181, 184, 185, 187, 235, 240

MacKinnon J 1, 32, 41, 159, 175, 235
Marx K 25, 214, 216, 217, 221, 227, 228, 229, 233
McKenzie R D 2
Meadows D H 208
Mills E S 1, 4, 5, 25, 32, 33, 34, 36, 41, 45, 50, 55, 56, 58, 68, 69, 70, 81, 82, 83, 84, 124, 125, 129, 149, 158, 159, 173, 174, 203, 235, 237, 238, 240
Mirrlees J A 5, 25, 33, 37, 51, 52, 53, 54, 57, 58, 68, 111, 131, 132, 133, 134, 135, 139, 148, 151, 162, 239
Mogridge M J H 130, 243
Mohring H 2
Montesano A 5
Morrill R L 212
Moses L N 20, 32, 71, 80, 81, 83, 88, 213, 237, 242
Muth R F 1, 3, 6, 18, 19, 20, 21, 22, 24, 25, 34, 45, 88, 105, 106, 111, 127, 128, 158, 198, 243
Myrdal G 209

Niedercorn J H 84, 132, 158, 238
Nourse H G 122

Oates W E 142, 177, 180, 187, 240, 241
Ohls J C 38, 122
Oron Y 34, 37, 54, 57, 58, 63

Papageorgiou G J 32, 89, 95, 97, 107, 109, 111, 238, 239
Pareto V 54, 67, 122, 139
Park R E 2, 10, 222
Parr J B 143
Pines D 50, 52, 55, 63, 105, 124, 145, 238, 242
Pióro Z 229
Polinsky A M 37, 61, 64, 66

Rashevsky N 138, 139
Ratcliff R U 2
Rawls J 51
Richardson H W 1, 13, 92, 95, 112, 122, 135, 140, 147, 150, 184, 239, 242

Ridker R G 60, 61, 239
Riley J G 51, 54, 58, 105, 111
Ripper M 124
Rodwin L 14
Rose-Ackerman S 39, 119
Rubinfeld D L 64, 66

Sawers L 221, 222
Schelling T C 122
Schmenner R 83, 237
Schultz T W 147
Scott A D 6, 25, 168, 217
Segal D 72, 73, 74
·Senior M L 200
Shavell S 37, 61, 64
Shefer D 74
Sheshinski E 33, 58, 69, 70
Solow R M 5, 42, 50, 51, 55, 58, 62, 111, 158, 218, 235, 241, 242
Sraffa P 217, 218
Starrett D A 68, 134, 139
Stern N H 53
Steuart, Sir James 25
Stevens B H 37, 168, 172, 173
Stewart C T Jr 146
Strotz R H 4, 51, 57, 58, 63
Swan T W 127
Swanson J A 144

Thünen J H von 2, 6, 10, 17, 25, 80, 217, 218, 239
Tiebout C M 37, 142
Tinbergen J 139, 144, 145, 146, 149
Tisdell C 136
Tolley G S 64, 140

Varaiya P 6, 40, 124, 139
Vernon R 83
Vickrey W S 36, 47, 48, 55, 57, 58, 125

Wabe J S 151, 239
Webber M M 213
Weber A 81, 82
Weiss H K 148
Wheaton W C 111
Wheeler D 72
Wieand K F Jr 60
Wilkinson R K 150, 239
Williamson H F Jr 213
Wilson A G 37, 172, 181, 198, 200, 201, 202, 203, 240, 241
Wingo L 3, 6, 18, 22, 23, 24, 25
Wright C 63

Yamada H 106, 151

Zipf G K 149

Subject index

Agglomeration economies 72, 74, 179, 198, 219, 228, 238
  in CBD 68-72, 90-92, 93
  empirical studies of 72-74
  and firms 141, 147
  and social interaction 78-79
  and transportation 71, 90-91
Agglomeration potential 92
'Agora' model 75-78, 238
Agricultural land use 6-9, 18, 147
Air pollution 37, 60-64
Alternatives to NUE 168-212 *passim*, 240-241
Antecedents to NUE 6-30
Automobile industries 222

Baumol blight model 178
Benefits and costs of agglomeration 134
Bid-rent function 3, 14-18, 105-106
  atypical 102-104
  slope of 17
Boston 194, 195, 196, 198

Canada 181
Capital 136
  finance 220-221
Capitalism and cities 214-215, 225-229
Capitalization of land values 60, 61-62, 64, 72
CBD
  agglomeration economies in 68-72, 90-92, 93
  and interaction models 79-80
  land use 70-71
  substitutes 75-80
Central limit theorem 139
Central place models 147
CES production function 74
Chicago 19, 69, 80, 81-82, 83, 98
China 228, 229
Cities and Housing 18-22
Cities as public goods 75-80
Class conflict 164, 217-220, 224
Closed cities 131, 211
Coalitions and clubs 141-143
Cobb-Douglas function 20-21, 35, 42, 45, 54, 104, 112, 127, 150
Columbus 73
Compensating payments 140
Competition in land market 38-40, 217-220, 236
Complex systems 203-204, 211
Concentric zone model 8-9, 11, 13
Conflict theories of land use 217-220

Congestion 37, 44, 47, 50, 55, 57, 60, 69, 90-91, 125, 173, 219
  tolls 57-58, 166
Consumer durables 220-221, 222-223
Crest of metropolitan expansion 212
Cuba 228, 229
Cumulative disequilibrium 171-181, 241

Decentralization 130
  economics of 32, 98-101
  of employment 32, 67, 83, 86
Demand for housing 35, 46, 189-190, 192, 202
Density
  gradients 2, 21, 24, 79-80, 83-84, 148, 160, 164-165, 237, 239
  comparison with standard model 164-165
  discontinuous 161-164, 166, 167
  dynamics of 129-130
  employment 81, 82-84
  gamma 87-88
  and planning 166
  policy implications of 165-166
  income and rents 158-161
  low 152-156
Dependency theory 227
Detroit 186, 187, 188
Developing countries 102
Diffusion wave analysis 212
Discontinuous
  densities 161-167
  development 124-125
  rents 126-127, 177
Discrete vs continuous models 124-127
Distance and optimum geography 146-148
Distribution of city sizes 131-149 *passim*
Dow Building Cost Calculator 192
Dual 169, 170
Durability 124, 127-128
Dynamic urban model 201-203
Dynamics 112, 124-130, 165, 242-243
  compromises in 124-127
  of density gradients 129-130
  and housing 127-129
  urban 203-212

Economic base model 181-182, 184, 185
Economies of scale 57-58, 68-74 *passim*, 133, 139-140, 184-185
  empirical evidence for 72-74
Edinburgh 13-14
Emission controls 63

Employment
  decentralized 32, 67, 83, 86
Engels
  on town vs country 227-279
  and urban problems 225-226
Entropy maximization 172, 199, 200
Environmental quality 98, 106-109, 111, 122, 151, 152 (see also Air pollution and Externalities)
Equal hierarchical population 143
Equilibrium
  absence of 187-188
  in agora model 77-78
  competitive vs optimum 4, 52-53, 56-57
  locational 16-17, 19-20, 40, 85, 109, 151-152
  long-run 40-41, 126-127, 198
Examples of NUE models
  Mills's 45-50
  simple 42-44
  variety of 110-111
Exchange and use values 228
Exclusive zoning 32, 80-82, 97, 224, 237
Expectations 124, 231
Externalities 37-38, 60-66, 83, 119, 142, 151, 162
  negative 132, 140, 147-148
Externality rent 150-152, 154
Externality zoning 122-123

Family size 67
Finance capital 220-221
Fiscal crisis 231
Fixed-point algorithms 175
Flight to suburbs 4, 180, 223
Freight costs 98-99, 100

Gammaville 54
General Motors 222
Generative vs parasitic cities 227
Gravity models 183, 185, 198-203
Greece 76, 78
Group preference models 119-122
Growth poles 229
Growth theory
  and housing 127-129
  and NUE 2, 235

Hausmannism 225
Hierarchy models 110, 132
  of distribution of city sizes 139, 144-146
  Löschian 93-95
  multicentric 95-97

Hinterland effect 147
Homeownership 226
Homogeneity 36-37
Households
  heterogeneous 67, 97-98, 104, 111, 119-122, 142
  homogeneous 36-37
Housing 19, 34-35, 45, 47, 186, 188, 207, 219
  demand 35, 46, 189-190, 192, 202
  dual, markets 167
  durable, models 127-129
  and growth theory 127-129
  and race 21-22, 119
  search 29-30
  supply 20-21, 46, 190-191, 192
Hoyt's sector model 11-14, 119
  critique of 14
  social status in 12-13
Hurd on urban economics 10-11

Income
  distribution 67, 219
  and location 20, 97, 102-103, 105-106, 112, 120, 154-156, 161, 177
Indifference curves 26-28
Infrastructure 124-220

Kinetic theory of gases 157-158, 159

Land use
  competition 38-40, 217-220, 236
  concentric zones 8-9, 11, 13
  conflict theories of 217-220
  planning model 181-185, 235-236, 240-241
  sector theory of 11-14
  von Thünen model 6-9, 217-218
  in transportation 33-34, 50, 54-58, 70-71
Land values 10-11, 72
  capitalization of 60, 61-62, 64, 72
  and zoning 122-123
Leisure time 66, 107, 109, 154-156
Linear programming 1, 124-125, 168-175, 191, 240
Location
  industrial 80-81, 82, 83, 93, 144
  of subcentres 95
Locational constants 95, 147
Locational equilibrium 16-17, 19-20, 40, 85, 109, 122, 151-152, 155

Subject index

Locational interdependence 40-41, 113-123
  and densities 113-117
  and group preference models 119-122
  and race 117-119
  and zoning 122-123
Lognormal distribution of city sizes 139, 141, 144
London 102
Los Angeles 81, 98, 222
Lowry model 181-185, 240-241

Macroeconomic model 194-195, 235
Manchester 86, 87, 88
Marxism 213-230 *passim*
  and capitalism 214-215
  and city 213-214
  and finance capital 220-221
  and land-use competition 217-220
  and rent 215-217
  and suburbanization 221-225
  and town vs country 227-229
Mass transit 58-60
Mathematical tools 41-42
Migration 73, 138, 141, 143, 196, 206, 210-211
MIT econometric model 194-198
  evaluation of 197-198
  long-run adjustment 198
  macroeconomic 194-195
  spatial allocation in 196-197
Monocentricity 15, 31-32, 68-88 *passim*, 101, 237
Monopoly 215-216, 220
  capitalism 228, 230
Multicentric city 32, 72, 89-101 *passim*, 131, 187, 237-238
  and linear programming 174-175
Multiplicative model 139

NBER model 185-193
  compared with NUE 193
Negative exponential gradient 79, 84-85
Neighbourhood preference premiums 120-121, 122
New Towns 229
New Units of Settlement (NUSs) 214
New York 73, 102
Nonspatial models 127-129, 211-212
Numerical analysis 48-49, 55, 125

One-dimensional city 31
Open city models 65-66, 140
  vs closed 130
Operationality of NUE 234-243

Optimal city size 69, 75, 136-138
Optimum geography 131-149
  coalitions and clubs model of 141-143
  equal utility in 140-141
  guidelines 148-149
  hierarchy models 144-146
  and location 144
  marginal product model of 138-139
  Mirrlees model of 132-136
  space and 146-148
  and Tiebout model 137, 142
  variable production functions in 136-138
Origins of NUE 1-5

Pacific Electric 222
Paris 130, 145
Penn-Jersey model 168-171
'Piazza' model 79-80
Pittsburgh 73, 186, 187, 188, 193
Policy implications
  of NUE 234, 239-240, 243
  in *Unheavenly City* 232-233
  in *Urban Dynamics* 208-209, 211
Political economy 213-233 *passim*
Positive feedbacks 177, 203, 209
Poverty, subculture of 231-232
Preferences 104, 111, 119-122
Primal 169, 170
Primary circuit 214-215, 230
Production
  neglect of 33-34, 70
Production functions
  and agglomeration economies 68-69, 72-74, 132
  and optimal city size 136-138
Property taxes 66
Public goods 37-38, 64-66, 75-76, 142
Public sector 37-38, 64-66

Racial discrimination 117-119
Range 93
Rank size rule 148
Rent 2, 10, 25, 98, 100, 130, 158-159, 160
  absolute 216, 218, 219
  and decentralization 98-101
  externality 150-154
  gradient 7, 12, 17, 24, 43, 52, 66, 84-85, 105, 112, 126, 238-239
  positive 23, 150-157, 238-239
  and race 117-119
  and housing 127-129

Rent (continued)
  and Marx 215-217
  maximization 169-170
  minimization 169-170
  scarcity 218
  and sector model 12-14
  in von Thünen model 6-9
Residence-workplace model 198-201, 202, 203
Residential attractiveness 109, 111-112
Rural vs urban life 132-134, 227-229

Sandwich method 176
Satisficing 29-30
Secondary centres 89-90, 92 (see also Decentralization)
Secondary circuit 214-215, 230
Sector theory 11-14, 119
Sexism 224-225
'Sharing-and-interaction' 75
Simplicial search algorithms 175-177
Simulation models 185-198, 203-212, 241-242
Slums 22
Social status and land use 12-13
Spatial distribution matrix 183
Spatial interaction models 198-203
Specialization 99, 146
Sperner's Lemma 176
Sraffa model 217-218
Studies in the Structure of the Urban Economy 45-50
Suburbanization 59, 63, 130
  and Marx 221-225
Subways 58-60
Supply of housing 20-21, 46, 190-191, 192
Surplus value 215

Taxes 64, 207
Tellow 6
Threshold 93
von Thünen model 6-9, 217-218
Tiebout model 37, 142
Time
  allocation of 76-77, 106-109
  leisure 66, 107, 109, 154-156
Town vs country 227-229
Trade-off model 119-120, 161

Transport orientation 82
Transportation 35-36, 149, 165, 219
  advantages 90-91
  in circular city 56-57
  expenditure function 85-88
  land-use allocation 50, 54-58
  and land-use model (Wingo) 22-25
  and linear programming 173-174
  mass 58-60
  technology 56, 136
  (see also Congestion)
Travel costs 44, 87, 98, 153
Unheavenly City 230-233
Urban crisis 4, 231
Urban dynamics 203-212, 241
  critique of 209-212
  multipliers in 206-207
  nonspatial character of 211-212
  and policy 208-209
  sectors in 205-206
Urban growth model 179-180, 206-208
Urban renewal 208, 225-226, 232, 233
User charges 230
Utility
  constraints on 30
  and distribution of city sizes 140-141
  indirect, function 65
  marginal 29, 105, 154
  maximization 1, 3, 15-16, 26-30, 41, 42-43, 62, 64-66, 76, 78, 95-96, 107, 118, 142-143, 150, 152-153, 154
    assumptions behind 51-54
    in NUE 51-54
  theory explained 26-30

Vintage models 127-129

van der Waals equation 158, 159, 162
Wages 6, 64, 107, 154
  and city size 64
  gradient 20, 84, 88, 156
Weberian location theory 81-82

Zoning 39-40, 166, 184
  exclusive 32, 80-82, 97, 224, 237
  externality 122-123
  fiscal 123
  and land values 122-123

For Product Safety Concerns and Information please contact our EU
representative GPSR@taylorandfrancis.com
Taylor & Francis Verlag GmbH, Kaufingerstraße 24, 80331 München, Germany

www.ingramcontent.com/pod-product-compliance
Lightning Source LLC
Chambersburg PA
CBHW062122300426
44115CB00012BA/1771